Marcin Słoma

Additive Manufacturing of Structural Electronics

T0074414

Also of Interest

3D Printing with Light
Edited by: Pu Xiao und Jing Zhang, 2021
ISBN 978-3-11-056947-6, e-ISBN (PDF) 978-3-11-057058-8

Additive Manufacturing
Science and Technology
Emrah Celik, 2020
ISBN 978-1-5015-1877-5; e-ISBN 978-1-5015-1878-2

Rapid Prototyping, Rapid Tooling and Reverse Engineering
From Biological Models to 3D Bioprinters
Kaushik Kumar, Divya Zindani and J. Paulo Davim, 2020
ISBN: 978-3-11-066324-2; e-ISBN 978-3-11-066490-4

Smart and Functional Textiles
Edited by: Bapan Adak and Samrat Mukhopadhyay, 2023
ISBN 978-3-11-075972-3, e-ISBN (PDF) 978-3-11-075974-7

Electrochemical Methods for the Micro- and Nanoscale
Theoretical Essentials, Instrumentation and Methods for Applications in MEMS
and Nanotechnology
Jochen Kieninger, 2022
ISBN 978-3-11-064974-1, e-ISBN (PDF) 978-3-11-064975-8

Marcin Słoma

Additive Manufacturing of Structural Electronics

—

DE GRUYTER

Author
Prof. Marcin "Polnicki" Słoma
Warsaw University of Technology
8 Sw. Andrzeja Boboli St.
02-525 Warsaw
Poland
marcin.sloma@pw.edu.pl

ISBN 978-3-11-079359-8
e-ISBN (PDF) 978-3-11-079360-4
e-ISBN (EPUB) 978-3-11-079366-6

Library of Congress Control Number: 2024930291

Bibliographic information published by the Deutsche Nationalbibliothek
The Deutsche Nationalbibliothek lists this publication in the Deutsche Nationalbibliografie;
detailed bibliographic data are available on the Internet at http://dnb.dnb.de.

© 2024 Walter de Gruyter GmbH, Berlin/Boston
Cover image: Marcin "Polnicki" Słoma
Typesetting: VTeX UAB, Lithuania
Printing and binding: CPI books GmbH, Leck

www.degruyter.com

Kochanej Żonie

Preface

The book presents the most important developments to date in the field of additive manufacturing of electronics in its broadest sense, using various functional materials and techniques. The elements presented here are organized in a workflow starting with a description of the additive manufacturing approach, an interesting and quickly developing printed electronic technology, an alternative approach to fabrication of structural electronic circuits, adaptation and application of various materials for additive manufacturing of electronics and ending with challenges and the perspective being in front of presented here domain. Of course, the main and largest part of the book deals with examples of additive approaches to the fabrication of structural electronic components and circuits, covering the scope of conductive materials and basic passive elements, active components with semiconductor materials, photonic components and systems, energy storage devices, a vast area of sensors and micro-electromechanical systems. All this fulfills the book's main goal of presenting potential solutions and possible practical applications of additively fabricated structures for electronics.

My intention behind the preparation of such a book was to deliver a thematically concise and highly descriptive work, being far from a formalized encyclopedic description, and at the same time, based on a large number of scientific sources for further exploration by readers. Here, I have focused on presenting global developments in the research and application of various types of materials, including composites and very popular nanomaterials in the adaptation for 3D-printed electronics. Many representative scientific literature reports, as well as selected industrial research and development approaches, are presented here to paint the picture of the current knowledge in this field as fully as possible. As there are many research studies on the electrical and mechanical properties of materials and structures tailored for additive manufacturing, including a vast range of nanomaterials, I have placed particular emphasis on the less detailed but more interesting from a practical point of view, topic of possible applications in the field of electronics and related areas such as biomedical applications, new materials or eventually "4D printing" for the new forms of manufacturing.

The primary objective of this book is to analyze the development of additive manufacturing, electronics technology and printed electronics, which are synergized in a new direction for research and development called 3D printed structural electronics. All these areas have a significant impact on the development of modern technology and industry. Therefore, it is important to introduce the reader to all of these topics before presenting the ultimate goal of additive manufacturing of electronics in detail. The need to address the topic of additive structural electronics emerges from the huge and growing interest observed in the wave of reports on practical or potential applications of 3D printing, nanomaterials or printed electronics. While this is, in fact, related to the high possibility of utilizing remarkable electronic, mechanical, thermal or optical properties of new materials, most of this information is related to studies published at the initial stages, later distributed by media, suggesting almost unlimited application possibilities,

https://doi.org/10.1515/9783110793604-201

sparking the imagination of applications that could revolutionize our world of technology or science. Readers need to be aware that we will encounter a lot of problems when trying to exploit these unique properties presented in scientific studies to be routinely applied in large-scale applications in the industry. After all, with the development of civilization, we have learned to introduce new techniques for synthesizing materials, selecting their properties, building structured and predictable macrostructures, and we have learned to scale up processes for massive industrial solutions. The same will be true with the additive structural electronics, but it will take some more time than presented in the media. In addition to existing research directions and applications, the book's last section presents recent research results in the field related to additive manufacturing, stimulating potential applications of 3D-printed structural electronics in emerging technologies. By doing so, I wanted to present developments for future electronic applications and other industry implementation branches, as well as articulate research and technological areas where we can look for new discoveries and breakthroughs.

The book should be understandable both by graduate students with knowledge of mathematics, physics, chemistry and materials engineering, but it should also be useful to researchers from other research areas than those related to mechanics, electronics or materials engineering, who are looking for interdisciplinary inspirations to the development of new research directions in their fields. This goal will be achieved if, after reading the book, the reader changes their perception of additive manufacturing and looks at it as a field that concerns all research aspects of the modern world. While the topic described here is a new paradigm shift, still in the initial development stage, this book is to shed some light on that area of interest and present the origins, available tools and directions to explore. All this is an element of current state-of-the-art achievements described in scientific literature and emerging solutions or profiles of the companies entering the market, with all related challenges and perspectives. Readers familiar with electronic technology will see the new directions for the development, and those familiar with the printed electronics approach will see the broader picture of additive manufacturing and possible new ways for advancement, while experts in additive manufacturing will have a chance to explore new possible directions of their matured techniques. This goal is not necessarily achieved in the course of reading the book "from cover to cover." The individual chapters have been prepared in such a way that they can be read selectively and used as a reference for further studies on the subject. In this aspect, it can also be used by teaching staff as a textbook for classes on new electronic technologies or as an example of the application of materials engineering for additive manufacturing. In the text, I have included references to numerous literature studies touching more deeply on the topics covered, which are recommended for readers who wish to learn more details than could be included in this study. This is due to my conviction that it was impossible to include descriptions that would satisfy all readers. This is also related to the specificity of the presented research areas, which are open chapters in the research aspect, and many of the recent studies produced during the development and publication of this book were not included in this book. This study is based on my experience gained during the last decade of research in the field of additive manufacturing, nanomaterials and printed electronics.

Acknowledgments

This work was supported by the Foundation for Polish Science, within the project "Functional heterophase materials for structural electronics" (First TEAM/2016-1/7), cofinanced by the European Union under the European Regional Development Fund; supported by the Institute of Metrology and Biomedical Engineering, Faculty of Mechatronics at the Warsaw University of Technology.

https://doi.org/10.1515/9783110793604-202

Contents

1 Introduction

Electronic devices are all around us these days and are an integral element of our daily routines as assisting systems for processing information. They have become one of the essential necessities of life in modern society, from entertainment to healthcare, from military equipment to toys, from household appliances to spaceships. Over the course of the last decades, electronic devices, such as computers and smartphones, revolutionized the way we collaborate, communicate, play or even live on a daily basis. Integrated with such rapidly evolving technology is the requirement of continuous development of materials, components, systems, and newer, more sophisticated and efficient manufacturing processes. Initially, electronics was mainly used in military and medical systems, later introduced for households and entertainment. With time, a great emphasis was put on the ever-increasing amount of information being processed. To meet these demands over decades, the main direction of electronics development was focused on increasing the performance of components and circuits. In its early years, the 60s, a pioneer in electronics development, Gordon Moore introduced a law predicting a continuous increase in the performance of electronics through an exponential increase in the number of transistors on an integrated circuit called Moore's law [1]. In the last decade, we have repeatedly approached the limit of this principle, and currently, the important limitation is reaching the physical limit of miniaturization (the size of atoms) and the increasing problem of dissipating the heat generated by integrated circuits [2]. This is particularly relevant today, as we can find electronics in a wide range of products. It is no longer just multimedia and communication devices such as televisions, computers or telephones but entire electronic modules that are present in most new household appliances, cars or even bicycles. In view of these phenomena, the proposed direction of developmental research in the field of electronics is the development of new forms of electronic devices, which would allow them to be implemented more easily in increasingly demanding applications. The result of this research is the emergence of a whole range of new branches of electronics such as elastic, printed, organic or, finally, 3D printed structural electronics. But let us start from the beginning.

The conventional electronics approach, namely the most of surrounding us electronic devices, consists of electrical and electronic components and circuits fabricated by so-called conventional electronics manufacturing methods. These manufacturing methods are highly complex in their procedures, comprised of a series of subtractive (rarely additive) processes. The most commonly used are photolithography, etching, masking, physical and chemical vapor deposition or laser ablation [3, 4]. Such production procedures involve most often expensive clean rooms and well-controlled conditions. The most basic element of almost all electronic devices is the Printed Circuit Board (PCB). It is one of the most essential components in any electronic device or system, being a corresponding electrical circuitry, which is used as a base element for mounting and connecting all other electrical components by soldering or other means of connection (adhesives, bonding, etc.). The conductive paths are fabricated

https://doi.org/10.1515/9783110793604-001

Figure 1.1: PCB circuit with mounted electronic components.

from highly conductive metals, most often from copper thin foils laminated onto the dielectric substrate, usually fabricated from a composite material such as epoxy-glass (FR-4) or phenolic-paper (FR-2), being the most popular and commonly used base substrates for PCBs due to their low cost, mechanical, electrical and thermal properties. Having conductive paths on dielectric substrates, we can distinguish the main types of circuits: single-layer, double-layer and multilayer PCBs, varying with the number and orientation of conductive paths. With more layers in the PCB circuits, we can significantly improve the packaging density of electrical components, and this increases the functionality of the system in the same space of the volume. Of course, this goes hand-in-hand with the higher price of multilayered PCBs. An exemplary electronic circuit board consisting of PCB and mounted components is presented in Figure 1.1.

This book is focused on the topic of 3D printed structural electronics, relating also to printed electronics technology. Here, we are dealing with the term printed circuit board, which is very misleading when put side-by-side with mentioned additive techniques. In the fabrication process of PCB, though the "printed" term is in the very name of these circuits, there is no printing or additive technique used as a crucial element of fabrication. As an example, the general fabrication process of a double-layer PCB is described as the following steps:

– copper foil lamination on the dielectric substrate,

- shape cutting of the final PCB and drilling of the holes and vias (machining processes),
- additional copper plating for conductive vias,
- photolithography patterning of paths and other elements (pads, antennas, etc.),
- chemical etching of the circuitry.

Additionally, there is also solder mask patterning and printing of information elements (the only printing process for PCBs), with additional cleaning, pretreatment and finishing steps in the middle of the whole process. As we can see, only information, such as marking, component placing and their names are related to the actual printing process, but that has very little to do with the main circuitry purpose. A more sophisticated process of semiconductor integrated circuits (IC) fabrication also involves many similar steps as masking, developing or etching. A few additive processes involve vacuum deposition of conductive paths or dielectric layers, but this is far from selective additive fabrication described in later sections of this book. For detailed information about the fabrication of PCBs and ICs, readers are advised to refer to the mentioned earlier handbooks or related manuscripts covering the fundamentals of electronics technology manufacturing.

But we are here not for the conventional electronics technology elaboration. And while this area of technology is continuously developing with more and more impressive achievements, the world of technology continues to advance, and one area that is showing tremendous potential for growth and innovation is the field of alternative electronics manufacturing. Here, additive techniques play a key role in innovation, enabling new formats of fabrication on elastic substrates, nonplanar surfaces or embedded in the construction elements, opening the possibilities for new design ideas, simplifying production processes and introducing personalized capabilities. Among the innovative approaches, fully 3D printed electronics is the most promising technology, allowing to fully integrate the electronic circuit into the device structure. In practice, this will allow a new approach to the way devices are manufactured and the development of a system that produces both the electronic and mechanical parts simultaneously, which is almost impossible for the current conventional approach. But to be fair, it needs to be mentioned that the idea of embedded or buried components is also explored for PCBs with passive and active components. Yet with the truly additive approach, this creates a lot more opportunities for higher integration and functionality of the systems, which will be presented in the further sections of this book.

2 Additive manufacturing

Additive Manufacturing (AM) is often used as an umbrella term describing a variety of novel manufacturing technologies enabling the fabrication of spatial objects. While it usually refers to the three-dimensional, hence the popular name "3D printing," it might be as well used for two-dimensional object fabrication procedures. In both cases, this is a "layer-based" approach incorporating joining materials instead of subtraction. The definition of the American Society for Testing and Materials (ASTM) in ISO/ASTM 52900:2015 states that this is "a process of joining materials to make objects from 3D model data, usually layer upon layer" [5]. This definition incorporates both the use of computer-aided design of the three-dimensional object (CAD model) with computer-controlled fabrication from the generated slicing code, resulting in a sequential deposition of material along the Z-axis [6]. While computer-aided manufacturing (CAM) is well known for subtractive techniques, 3D printing incorporates the achievements of computer numerical control (CNC) automated subtractive processes for building the elements by adding the material instead of removing it. The devices used in additive manufacturing are collectively referred to as 3D printers. Despite the many differences in the process details and parameters between the various types of 3D printing, the workflow to obtain the final product is often very similar. One important aspect of all additive techniques is the digital printing approach, also known in the printing industry for various types of inkjet printing. As mentioned previously, the first step is to create a three-dimensional CAD model of the physical object, prepare a CAM program for the individual steps and parameters of a 3D printer, and finally perform postprocessing operations to prepare the final product or improve its properties.

In most of the sources relating to the history of additive manufacturing, the origins are traced back to 1984 when Charles Hull filed a patent for a "system for generating three-dimensional objects," also giving it the name "stereolithography" (SLA) [7]. Charles later founded the first 3D printing manufacturing company, called 3D Systems, which continues to manufacture 3D printing equipment. We could argue if this is really the dawn of additive manufacturing or if it is the most important step in 3D printing history. Probably the most known 3D printing technique is Fused Deposition Modeling (FDM) developed by Scott Crump in 1988, a building block for his company Stratasys, which is a market leader in additive techniques [8]. Also, development of the Selective Laser Sintering (SLS) technique in 1986 by the University of Texas student Carl Deckard [9] set the milestone for industrial applications of 3D printing in such techniques as Selective Laser Melting (SLM), Direct Metal Laser Sintering (DMLS) or Electron Beam Melting (EBM), a core technologies for high-quality and reliable metal parts, sometimes outperforming other known production methods. But even before Chuck Hull, there were inventors or visionaries of additive manufacturing such as Hideo Kodama describing in 1981 a laser beam curing process [10], David Jones describing in 1974 a SLA-like process in the sort story [11], Johannes Gottwald in 1971 filing a patent for a Liquid Metal Extruder [12], Stanisław Lem in 1955 describing a 3D printing process in one of his science

https://doi.org/10.1515/9783110793604-002

fiction novels [13] or other sci-fi writer, Murray Leinster, describing in 1945 a robotic arm for making a plastic object from pictures [14]. Finally, before printing in 3D dimensions, printing techniques were in use for centuries, providing books, newspapers or recently focusing on the packaging for logistics in the face of the digitalization of text. These printing techniques used in the printing industry, described in detail in later sections, also present an additive approach to the fabrication of goods.

Initially, additive techniques were developed as a rapid prototyping approach. They attracted attention at the beginning of this century when 3D printing became increasingly accessible thanks to the explosion of low-cost desktop 3D printers [15]. This attracted more attention to the area of additive manufacturing giving more traction for the development of various new printing technologies [16–19]. Thanks to that a new regions of professional applications in mechanical engineering emerged like aerospace or automotive [20–25], but also spreading beyond industry, to medicine for personalized implants and prostheses [26–28] drugs [29–32] or regenerative scaffolds [33–35] and biological tissues and organs printed from living cells [36–39].

Although additive manufacturing is still a relatively new technological approach, it is expected to revolutionize the manufacturing industry, building on the presented advantages over traditional manufacturing methods, and this way establishing a stable position on the market [40–43]. The forecasts on global market grow for additive manufacturing is at 15 % (CAGR, 2015–2025) exceeding $10 billion US dollars in 2021 [44]. The biggest players on the market are North America (42.2 %), followed by Europe (36.5 %) and Asia Pacific (11.5 %), with applications in automotive (40.5 %), aerospace (23.3 %), medical (18.2 %), consumer goods (13.6 %) and other (4.3 %) [45] but new application regions are rapidly emerging. This is due to the main advantages posed by additive manufacturing over conventional manufacturing techniques, including rapid prototyping reducing product development costs and time-to-market for new innovations, fast introduction of modifications and new designs without additional tooling costs or preparations, lower inventory costs or shorter supply chains, mass customization tailored for individual customer, small volume manufacturing is more cost effective but most of all the opportunity for almost limitless geometries and complex structures fabrication impacting significantly higher integration by reduce separate parts and assembly procedures, having the highest impact on the weight reduction and lowering production cost.

Thanks to the development of computerized manufacturing systems, new direct printing methods have been made possible. Such maskless techniques are often called digital printing because they do not require additional masks (printing forms, such as stencils, screens or templates) or any other specific tools to be fabricated. The design can be directly printed on the CNC setup, allowing a wide variety of designs to be applied almost instantly, requiring only the application of specific materials (such as inks in the printing industry). Such direct/digital printing techniques, thanks to their high degree of adaptability, allowed the advent of additive manufacturing. According to the mentioned ISO/ASTM 52900:2015 standard, the AM processes can be divided into seven unique categories: binder jetting, direct energy deposition, material extrusion, material

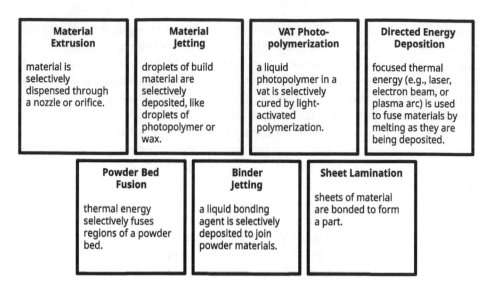

Figure 2.1: Classification of additive manufacturing processes according to ISO/ASTM 52900:2021 [46].

jetting, powder bed fusion, sheet lamination and vat photopolymerization, summarized and presented in Figure 2.1.

During the years of development, a number of 3D printing techniques appeared on the market, differing mainly in the principle of operation, the used materials, fabrication speed or other advantages or dedicated final application in the industry. At the moment, we are facing two basic 3D printing technologies: deposition from a nozzle (polymers, inks) and batch solidification (liquids, powders) with minor modifications (i. e., jetting the binder onto the powder). In the further sections, a detailed subdivision into several categories is presented, focusing on the method of layer deposition and used materials. Presented techniques will not be covered in detail, although this is not the main scope of this book. Readers could get familiar with the details of all additive and printing techniques in the vast resources of review papers and books covering this area of technology.

2.1 Fused deposition modeling

Fused Deposition Modeling (FDM is a brand name of the company Stratasys), also called Fused Filament Fabrication or Filament Freeform Fabrication (FFF term is used for open design platforms) is by far the most popular and widespread additive technique today. It is localized in a group of selective deposition techniques for melted material. The main principle of this technique is fairly simple, incorporating the extrusion of a thermally molten or semisolid material through a nozzle of normalized diameter, to form a single layer of the solidified, cooled material, and changing the nozzle position in the Z-axis by one step to repeat the process for the second layer [47–50]. The main elements of the printer are the 3D-axis kinematic system (sometimes 4- or 5-axis) [51, 52], the building plate (often heated) and the printing head consisting from extruder and heater [53]. The CNC microcontroller is used for controlling the movement of the kinematics, mainly the extruder head in the XY plane and the building plate in the Z-axis, and to control other printing parameters such as nozzle temperature, build plate temperature, feed rate of the material, cooling factor, etc. All these basic parameters (and many more) also need to be precisely synchronized.

The detailed description of the FDM process is more complicated, but as for all the following techniques it will not be covered here fully, and readers are advised to browse the literature for more information. The most important aspect to consider is the material deposition (Figure 2.2(a–d)). In the FDM process, the filament, most often a thermoplastic material, is fed from a spool to the electrically heated extruder and melted, transforming it from solid to highly viscous. Melted filament is mechanically pushed through the nozzle by pressure, induced from the fresh material transported by feeder rollers connected to the stepper motor. The extruded thermoplastic material is cooled shortly after deposition by ambient temperature or by additional fans located near the tip of the nozzle. This way material progressively adheres to the previously cured layers or to the build plate for the first layer. The movement for the printing head, the temperature of the heater, the feed rate of the filament at the rollers or the cooling speed are synchronized [54, 55]. This is a quite robust process principle, allowing the production of normalized thick layers, with fairly good adhesion to each other, is quite resistant to substrate irregularities, but if not properly controlled there is a high risk of delamination of the entire layer or element from the substrate, causing a defective part. It allows the fabrication of medium complex parts up to high-scale components (Figure 2.2(e,f)).

The XYZ kinematics of the FDM printer is not very different from the CNC milling machine. The heated nozzle is positioned in the XY plane to perform a preset profile trajectory, the same as for the milling spindle. In the same way, process resolution categories are defined: Z resolution and XY resolution, with additional minimum feature size [56]. Resolution on the Z-axis is directly related to the minimum layer height and should not be considered as a main factor determining the precision of manufacturing. In most of the 3D printing techniques, such as FDM, or later described direct ink write, VAT polymerization, powder-bed and jetting techniques, this value is correlated with the resolution of the motors and gears driving the printer nozzle or build plate. The same applies to the XY resolution determining the minimum horizontal movement of the nozzle, build plate or sometimes optical imaging system. This parameter, the XY resolution, determines the smallest horizontal feature allowable to be printed—a minimum feature size. Additionally, this is directly connected with the feature size determined by the diameter of the nozzle [18], and both of these factors should be taken into account when determining the precision of the fabrication with FDM. For most FDM equipment available on the market, the minimum feature size is in the range of 100 μm for polymers,

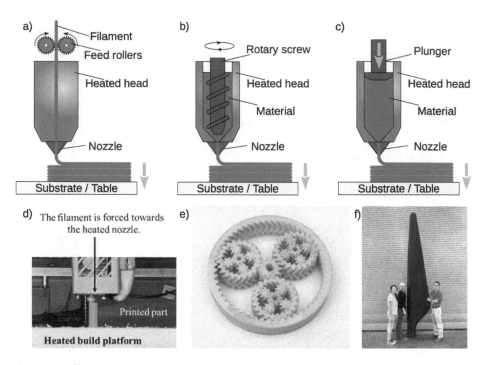

Figure 2.2: Different approaches for Fused Deposition Modeling additive manufacturing: (a) filament-based, (b) screw-based, (c) plunger-based. (d) Deposition of melted polymer in the additive fabrication process [53]. Reprinted from Springer Nature, under Creative Commons CC BY License, copyright 2023. (e) Example of a small size, high complexity part printed from polymer feedstock with the FDM technique. (f) 3D printed wind turbine blade with Big Area Additive Manufacturing [63]. Reprinted with permission from Solid Freeform Fabrication Symposium.

but for composites, it can be more than 200 μm due to the high viscosity of the plasticized composites [54]. The printing speed parameter is here defined by the length of the filament extruded via the nozzle, usually yielding 1–10 m/min [57–59].

By the definition, the leading materials used for this technique are various thermoplastic polymers, prone to thermal melting [59, 60]. In most cases, the thermoplastic filament has to be prepared beforehand using other extrusion processes, with respect to a standard diameter suitable for the printer's head and heater, but there are available FDM printers on the market, equipped with the hopper and screw extruder known from the injection molding process, allowing direct extrusion of thermally melted polymer pellets [61–63]. The group of polymers required for FDM processing are almost limitless. The most popular group of materials are acrylonitrile butadiene styrene copolymers (ABS), polylactic acid (PLA), polyurethanes (PU), polyamides (PA), polyphenylsulfone, poly(ether ether ketone) and poly(ether imide) [59, 64]. While it was originally developed to rapidly produce prototypes from polymeric materials, in the course of its development, polymer matrix composites have been also introduced, also for the fabrication of electrically conductive structures from carbon or metal composites [65, 66].

The main advantages of the FDM technique are the very simple construction and control of the equipment, which was the main driver for the introduction of inexpensive 3D printers, costing as little as a few hundred dollars, but also the availability of inexpensive and accessible polymer filaments and relatively high fabrication speed in the group of additive manufacturing techniques [15, 59, 67]. Some disadvantages are related to the need of fabricating supporting structures required for free-standing models that are not needed in some powder bed systems, relatively low printing resolution if we want to incorporate it for production of precise electronic devices and high possibility of nozzle clogging for high viscosity composites. However, the greatest advantage of this method in the scope of electronics printing is easy integration to achieve multimaterial printing [48, 68] by adding additional printing nozzles according to the number of processed materials. Moreover, in many cases, it would be easy to adapt commercially available devices for this application.

2.2 Direct ink writing

Another additive technique based on material extrusion is Direct Ink Writing (DIW). This technique is very similar to the above mentioned FDM printing and incorporates many common technical solutions in the construction of 3D printers. It also involves the continuous extrusion of material from a nozzle moving just above the surface of the substrate, and deposits sequentially layers on top of previous ones. The main difference here is that the printing material is a viscous functional ink instead of a solid polymer filament. This also eliminates the need for thermal melting of the material, thus additional heating is possible. The functional viscous ink stored in the container, often in the form of a syringe, is extruded via a needle-like nozzle with the use of pneumatic pressure, allowing the deposition on a substrate or previous layer [69–72].

Similarly, as for FDM, the DIW printer consists of XYZ kinematics that are CNC controlled. The work principle is also very similar, and the nozzle is positioned in the XY plane to perform a preset profile trajectory, the same as for the extruder in FDM. Here is the main difference between these techniques, while the system includes of the ink reservoir and the nozzle (glass, metal or polymer capillary) for the extrusion of the viscous ink. Such setup, similar to FDM, needs precise control and correlation of the pneumatic pressure, which is attributed to the amount of ink pushed out of the nozzle, with the position and speed of the XYZ movements. Pressure is usually generated on the piston via a pneumatic gas pump or motor-driven screw. Such a combination enables the formation of complex and fine patterns with high resolution [73, 74]. Figure 2.3 illustrates the schematics of the DIW dispensing systems and examples of 3D printed elements. Here, a bit more explanation is needed, while such a description of the DIW system is well known for the equipment used for paste, sealants or adhesives deposition, which

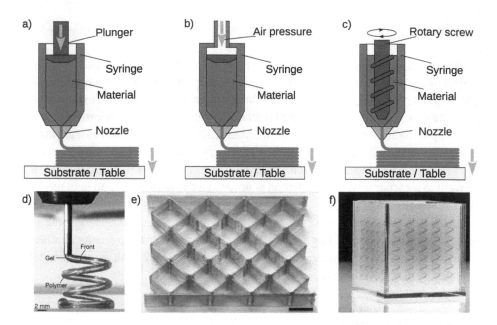

Figure 2.3: Different approaches for Direct Ink Write additive manufacturing: (a) plunger-based, (b) pneumatic-based, (c) screw-based. (d) Deposition of gel solution solidified after extrusion in the additive fabrication process [72]. Reprinted with permission from Springer Nature, copyright 2018. (e) Example of DIW printed triboelectric nanogenerator [74]. Reprinted with permission from John Wiley and Sons, copyright 2019. (f) Embedded 3D printing with DIW deposited viscoplastic materials into the gel matrix [88]. Reprinted with permission from American Chemical Society, copyright 2018.

are almost always deposited on flat surfaces or at most on curved substrates, but not building spatial 3D objects. In classical 2D deposition, extruded paste forms meniscus at the tip of the nozzle, and by wetting in the interaction with the substrate the ink attaches locally. With such sufficient wetting, printing is performed by drawing ink on the substrate by moving the nozzle in the XY plane and controlling the pressure applied to the ink, and the distance between the substrate and the nozzle. Such an approach is sufficient for one-layer deposition on the substrate. The challenge is to maintain the proper contact during printing on a curved substrate, but this is achieved by measuring the distance between nozzle and substrate (optical or contact measurements) or providing precise input of substrate topology to the equipment [75]. Going one step further, printing with DIW, a freeform 3D structure, besides the precise control of the nozzle position and dispensing parameters, also requires the special formulation of the ink that will maintain its structure after the ink extrusion out of the nozzle, without supporting substrate. For such attempts, an ink capable of maintaining the structure in the ambient air should be used thermally, UV or chemically activated or cured via solvent evaporation, or even laser sintered.

The main process parameters related to the XY and Z-axis resolution are similar to the FDM technique due to the mentioned similarities in the kinematics construction. Standard DIW equipment is capable of printing with the minimum feature size in the range very wide range from 500 µm down to even 1 µm [76], and also high aspect ratio structures up to ×1000 are possible with hybrid technique [77]. Besides the nozzle diameter, this parameter is heavily ink dependent [71, 78, 79]. The width of the lines depends mainly on the inner diameter of the nozzle but can also be modified by controlling other parameters such as material extrusion rate, nozzle movement speed and nozzle-to-substrate distance [80, 81]. Also, as the nozzle size decreases, the thickness of a single layer also decreases, which creates a challenge to maintain a proper distance between the nozzle and the substrate that should not vary by more than 0.5 of the inner diameter of the nozzle. The printing speed here is defined by the volume of extruded paste, usually in the range of 100–1000 mL/h [71, 82, 83].

The flexibility of this method lies in the wide availability of the materials to be applied. This also creates more opportunities for the size of printed features than with the FDM technique. The materials used for DIW are in the form of pastes or inks. In order to achieve very narrow lines, it is important to note that the applied material must have the right rheology. By modifying this parameter, the ink will flow easily through the nozzle or a paste will require a much larger driving force and shear-thinning properties. Most often the rheology of the ink can be controlled with the particles filling and its size, while solid particles should be around 10–100 times smaller than the inner diameter of the nozzle to prevent clogging [84]. To maintain its printed structure, the viscosity of the ink should be precisely controlled, depending on the feature size [85, 86]. When using low-viscosity ink, a postprocessing curing system is often used (such as UV light source) to cure extruded material instantly to retain its 3D structure [18, 71], or incorporating additional shear thinning or high viscosity gels, as a supporting matrix in external reservoir, and performing injection of low viscosity filler material from the nozzle directly into supporting matrix [87, 88]. Generally, DIW is a highly versatile technique in terms of applied material, and it is easy to prepare simple polymeric and colloidal suspensions in the form of inks or even adapt pastes from other printing techniques such as screen-printing [81]. It is all about the viscosity, and the range value is large (10^2–10^8 [cP]) [71, 89–91].

Direct ink write technique offers many benefits, including printing at room temperature and mentioned versatility of materials in the form of various suspensions, shear-thinning fluids, gels, foams, composite mixtures and even biological inks with living cells for easy fabrication of free-standing structures [92–98]. Some disadvantages are observed for very fine structures (down to micrometre range) related to complex ink formulation [18]. Like FDM, DIW also offers a major advantage for 3D printed electronics related to the simple implementation of multimaterial printing with multiple nozzles and ink reservoirs, allowing the integration of various functionalities in a one 3D printing format [99–101].

2.3 Direct ink jetting

Inkjet printing is a mature printing technique widely used in many applications over the past few decades. For sure, it was developed before anyone even thought about 3D printing, but as mentioned previously all printing techniques are additive techniques and, therefore, such mature technique as inkjet is presented here along with other 3D printing techniques. Inkjet printing most of all is well known from home and office desktop printers, the most prominent applications for digital printing of images and text onto paper substrates. In the printing industry and packaging applications, inkjet printing accounts for about 60 % of the direct printing market [102]. This is also related to the main principle of operation of this technique, which is fairly simple and quite similar to the previously mentioned DIW technique. Inkjet printing involves the formation and deposition of small droplets by a print head positioned just above the material. The desired pattern is created by moving the printing head so that the droplets overlap and form a consistent and uniform shape [103–107]. Because tiny droplets are formulated, low viscosity inks are required, typically containing pigment particles, a polymer binder and a solvent that evaporates after printing, leaving a thin layer of material on the substrate. Figure 2.4 demonstrates the working principle of the Multijet 3D printing technique and a few examples of elements fabricated with this approach.

The detailed working principle of the inkjet technique is much more complicated, mostly due to the several methods of droplet formulation and deposition approaches. Two main deposition methods for inkjet printing are continuous inkjet printing (CIJ) and drop-on-demand (DoD) inkjet printing [108, 109]. The continuous method is popular for high throughput and low-quality printouts and finds little attention in additive manufacturing and printed electronics. Here, a constant stream of ink droplets is ejected from the nozzle, and passing through an electric field charged droplets are deflected toward the substrates to create printouts. Remaining droplets are captured by a gutter and forwarded to the ink container for reuse [108–111]. The more popular, drop-on-demand printing method only ejects an ink droplet upon demand from the multinozzle printing head. Here, several droplet formulation techniques are also introduced, based on the actuator system in the printing head. The most popular and most conventional are thermal and piezoelectric systems [108, 112]. For both techniques, thermal and piezo, the main working principle is similar, as the actuation mechanism generates a pressure pulse pushing out ink droplet from the nozzle [108, 109, 113–115]. Thermal printers heat the ink in microseconds, which creates a thermal expansion of the ink, while piezoelectric printers apply voltage to a piezoelectric membrane, changing the pressure inside the ink chamber. In both cases, this propels and ejects a droplet from the nozzle. The shape and size of the droplet can be altered using pulse parameters (e. g., build-up rate, duration, frequency).

Here, the Multijet or Polyjet technique enters the stage of 3D printing. At first glance, the inkjet method, ejecting small droplets and forming thin layers of ink on the substrate

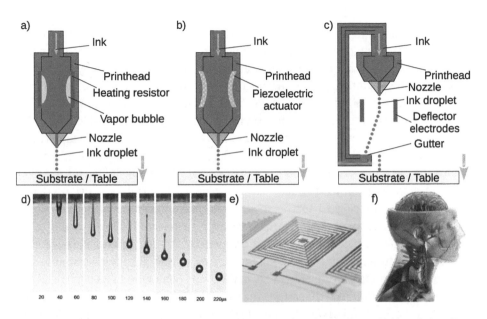

Figure 2.4: Different approaches for Direct Inkjet additive manufacturing: (a) thermal inkjet printing, (b) piezoelectric inkjet printing, (c) continuous inkjet printing. (d) Representative photo sequence of drop formation in the inkjet deposition process [115]. Reprinted with permission from American Chemical Society, copyright 2009 (e) Inkjet printed multilayered conductive coils for inductive sensing applications [122]. Reprinted with permission from Springer Nature, copyright 2017. (f) Vascular anatomical model fabricated with the Polyjet additive technique. Reprinted from www.stratasys.com with permission from Stratasys LTD.

does not seem like a natural technique for fabricating freestanding, spatial 3D structures. Yet, by applying the resin by printing heads moving above the working platform and instantly curing deposited materials using a UV light source, we are able to obtain quite easily a finished 2D layer, which brings us to the origins of 3D printing. Only the next step of lowering the platform is needed, and the process repeats [116, 117]. The only thing that needs to be modified is the additional printing head to apply a support material that differs from the main material and can be later removed (dissolved).

The inkjet technique requires very low viscosity inks, most often <0.25 Pa·s [116, 118, 119] for obtaining high-quality printouts and to prevent nozzle clogging [120], but with the use of inks with nanomaterials, it is fairly easy to fabricate printed electronic components by depositing functional and electrically conductive materials [109, 121, 122]. In these techniques, the droplet thickness determines the Z resolution of the printed elements. Ink droplets as small as 2–12 pL can be deposited, resulting in a few micrometers layer or even submicron dimensions [117, 123]. As it was for the FDM and DIW techniques, the XY resolution corresponds to the movement of the printhead, additionally determined by the footprint of the deposited ink droplet, resulting in the resolutions in the range of 5–100 µm [18, 116, 124]. The printing speed is defined by the volume of deposited ink and it is in the range of 100–1000 mL/h [116].

The main advantages of inkjet techniques are the use of various substrates, limited waste, easy adaptation of low-viscosity inks and high printing resolution compared to other techniques, crucial for high-integration electronic devices. Initially, the biggest drawback of DoD printing was its low efficiency, but this problem was solved by multiplying the number of nozzles in a single head, up to several thousand nozzles (fabricated with Microelectromechanical System technology (MEMS)), capable of printing with 1200 dpi a resolution and more [113, 125]. The biggest drawback is the high risk of nozzle clogging for highly loaded inks. One of the serious drawbacks of the inkjet technique is the use of low-viscosity inks to avoid nozzle clogging [120], limited for AJP [126]. While for graphics printing it is not a major problem, and printers are continuously working with several clogged nozzles, in electronics manufacturing such a situation is unacceptable, as less conductive material can lead to a deterioration in the quality of the transmitted signals, and in extreme cases to an open circuit. Another problem is the need to maintain a constant distance between the nozzles and the substrate/printout, while distance directly affects the kinetic energy of the droplet on impact, and might lead to a "coffee-ring effect" [127, 128] that affects the submicron quality of the printed features. Similarly, as for FDM and DIW techniques, inkjet printing is one of the most widespread in printed electronics, expanding in the 3D printed electronics applications, thanks to the easy adaption of multiple print heads for multimaterial deposition [116, 119].

2.4 Aerosol jet printing

The last mentioned direct deposition technique is Aerosol Jet Printing (AJP). This is also an ink-based technique, but compared with inkjet, AJP is a more recent technology, developed as a response to some drawbacks of inkjet systems, especially regarding printing on nonflat substrates. It shares many similarities with previously mentioned direct-write techniques, except the material deposition, which relies on aerosol deposition instead of defined volume droplets. The aerosol-forming system is the heart of the AJP printing technique and needs more explanation. In general, low-viscosity inks are converted into an aerosol, later focused into a stream, and ejected with high speed from a nozzle [129, 130]. First, an aerosol is formed in an atomizer chamber that converts the liquid ink into aerosol droplets of micrometer diameters. This process can be performed with two quite different approaches: pneumatic atomization with high-flow gas or ultrasonic atomization with a piezoelectric element. Atomized ink in the form of aerosol mist is transported to the deposition head by a carrier gas and later formed into a narrow stream by cylindrically surrounding sheath gas in the print nozzle. Such aerodynamic interaction between the sheath gas and the aerosol stream efficiently formulates a sharp, focused aerosol stream. Such a procedure of transporting aerosol in carrier gas and later surrounding it with shield gas while maintaining a laminar gas flow is quite complicated, but allows efficient formulation of the focused beam of the aerosol jet in the order of 10 μm, and also protects the nozzle from clogging. Figure 2.5 shows the working

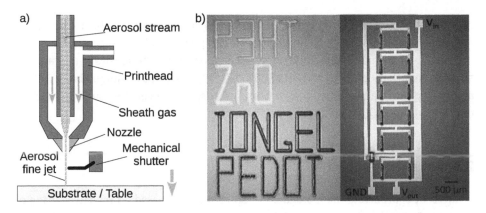

Figure 2.5: (a) Schematic representation of working principle of the Aerosol Jet Printing process. (b) Photograph of AJP printed functional inks (P3HT, ZnO, ion-gel and PEDOT:PSS) and image of a single complementary inverter, and a 5-stage ring oscillator, also AJP printed [139]. Reprinted with permission from John Wiley and Sons, copyright 2014.

principles of the AJP deposition process, and an example of fine patterns printed with this technique.

Aerosol Jet Printing is also an ink-based technique, and thus requires low-viscosity inks, allowing printing solutions and dispersions with a viscosity in the range from 1 to 1000 cp [130, 131]. Here, we have more flexibility with materials selection than for inkjet printing, due to the efficient use of two atomizing methods (pneumatic or ultrasonic): the first one is more efficient for high viscosity inks and the second one for low viscosity inks, respectively [130, 132]. More options for ink compositions enables the use of a wide range of functional materials such as metal nanoparticles [133], carbon-nanomaterials [134], ceramics [135] and even biological material [136].

The XY resolution is determined by the kinematics construction, which is often similar to the positioning system of FDM or DIW techniques, although very popular adaptations of AJP include 6-axis positioning of the substrate. The minimum line density and width (feature size) is the diameter of the aerodynamically focused stream—down to 10 μm. Layer thickness (Z resolution) is comparable to inkjet with a few hundred nanometers [130, 137]. Printing speed defined as the volume of deposited material is up to 1200 mL/h [138].

AJP combines many advantages of the previously mentioned direct deposition techniques. One is the versatility of the materials used, even more than for the inkjet technique, and eliminated the problem of nozzle clogging and coffee-ring effect also observed for inkjet and application for various substrate geometries, maintaining the high resolution of print. The complexity of the system is a major drawback with the high cost of implementation. Also, the printed line morphology (width, thickness and roughness) is highly dependent on the precise control of the process parameters, such as atomization settings, sheath and carrier gas flow rate [131, 137]. But the most powerful advantage of AJP in the scope of application for 3D printed electronics, not possible for any other mentioned technique, is the ability to print on complex, nonplanar surfaces, due to the high stand-off distance (up to 5 mm) between the nozzle and the substrate, tolerating to any nonuniformity of the surface [130, 133, 137, 139]. Also, this technique has a unique ability to print in all directions, including upwards, and on complex surfaces, such as orthogonal sides of a cube [137] or a golf ball [140].

2.5 Vat polymerization

A different approach to the fabrication of 3D elements is based on selective curing or binding of materials, instead of selective deposition. As already mentioned, the first additive manufacturing technique on the market was stereolithography, utilizing a laser photocuring of a special reactive liquid polymer layer-by-layer. The vast group of photocuring additive techniques is called Vat photopolymerization, due to the use of liquid resin in a vat or other form of vessel. Vat techniques are not commonly used in the fabrication of 3D electronics, which is mostly related to the problematic use of conductive materials. But this technique is one of the key players in the additive manufacturing market and, therefore, needs to be mentioned here along with other 3D printing approaches. What is more, recently new perspectives have emerged for the application of photocuring techniques for the fabrication of conductive structures, which will be briefly presented in later sections.

In general, in photopolymerization liquid monomers/oligomers are used that can be cured/polymerized when exposed to a light source of a specific wavelength. The reaction requires a material called a photoinitiator, which produces radicals when exposed to light irradiation, and reacting with monomers/oligomers initiates polymer chain growth [141]. The most common light source for photocuring is UV (<400 nm), but a range of other light wavelengths is not limited and may be selected for specific initiators included in the resin. This is a general description of the reaction utilized in vat polymerization additive manufacturing, but each technique varies with different use of light distribution or additional process parameters. Regardless of the differences described in this section, in further sections of this book, I will refer to the whole group of photocurable resin-based techniques as VAT or SLA, to simplify the description.

The two most common approaches for VAT polymerization additive techniques are Stereolithography (SLA) and Digital Light Projection stereolithography (DLP) with the latter modified also to Liquid Crystal Display stereolithography (LCD). SLA utilizes a laser beam as the light source solidifying the resin by scanning the corresponding cross-section area of the current layer of the material with a focusing system. On the level of one layer, this process is almost identical to the photolithography procedure used for the fabrication of electronic printed circuit boards (PCBs). Next, the platform moves one step by the height of the layer (Z step) and the process repeats. The entire layer consists of volume elements (voxel) created one at a time [7, 142]. Additionally, a support structure is needed to efficiently fabricate a solidified 3D model, which requires an additional post-processing operation of trimming supports after the printing process. Due to the need to scan with the optical system the entire area of the layer, which takes a considerable amount of time, SLA suffers from relatively low printing efficiency. The quality of the beam focus also affects the accuracy and precision of the print. Such problems with speed and focusing are eliminated in the modification of the SLA technique, a DLP [143–145]. This is a very similar technique, but while SLA illuminates one voxel or line at a time, in the DLP technique polymer is cured by exposing the entire layer at the same time with the projector [146]. A digital light processor (hence the abbreviation DLP) is a matrix of microsized mirrors, allowing each pixel to irradiate selected voxel on the surface of the polymer. A modification of this technique uses LCD projectors or masks, lowering the cost of the equipment, and allowing for comparable results [147]. Another significant modification of the VAT technique is Continuous Liquid Interface Production (CLIP) [148, 149] also utilizing the DLP approach, but the biggest difference is the continuous process of resin polymerization during the model protrusion form the resin, instead of step- by-step layer fabrication. This is achieved by incorporating an oxygen-permeable window, enabling a micrometer-scale layer of uncured resin (a dead zone) to exist between the window and the part. Along with the continuously projected images (a sort of a video instead of a set of slides), the part is steadily drawn as one print instead set of layers, allowing for more than 100× faster printing speed than conventional DLP [149]. Another modification of SLA, allowing the fabrication of micrometer scale components with submicrometer scale voxels is Two-Photon Polymerization (2PP) [150–152]. This technique also uses a laser source, ultrashort IR (infrared) femtosecond laser pulse, focused on the point in the resin basin. This allows to polymerize a tiny voxel within the resin and not on the surface, enabling 100 nm feature size, but at the same time is limited to a very low volume of $1\,\mathrm{cm}^3$ [16]. The last mentioned here is volumetric VAT polymerization, also utilizing similar principles to DLP printing, yet here an irradiating light creates a 3D object while rotating a vat [153, 154]. This approach is sometimes compared to reverse tomography, while the set of images is displayed with the projector into a transparent resin container, and this way a reconstructed object appears in polymerized resin. This is by far the fastest 3D printing technique, while the macroscale objects can be completed within 1 min [155]. The working principle of the two most common VAT polymerization techniques (SLA and DLP/LCD) and volumetric 3D printing schematic, with examples of fine detail structures fabricated with VAT techniques are presented in Figure 2.6.

For most typical VAT printing systems, such as SLA and DLP, the Z resolution is similar to FDM or DIW techniques, which is related to the resolution of the positioning system of the build plate. Additionally, the UV light penetration depth of about 100 μm from the surface of the resin, limits the resolution in the Z-axis [155, 156]. The minimum feature size in the XY plane is determined by the spot size of the laser for SLA (10–200 μm)

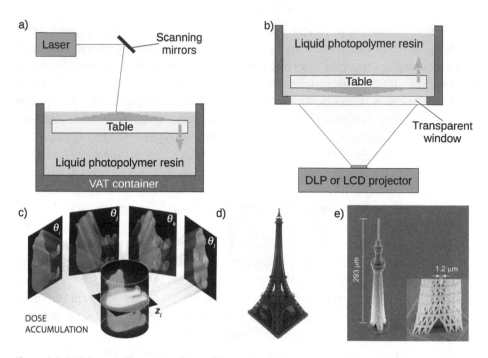

Figure 2.6: (a) Schematic illustration of stereolithography (SLA) process, and (b) alternative DLP approach. (c) Schematic illustration of volumetric additive manufacturing [153]. Reprinted with permission from Elsevier, copyright 2021. (d) Example of the fine detail structure fabricated with VAT photopolymerization approach [156]. Reprinted with permission from John Wiley and Sons, copyright 2019. (e) SEM image of "Tokyo Skytree" fabricated by two-photon polymerization (2PP) [150]. Reprinted with permission from Springer Nature, copyright 2013.

[142] and pixel size for DLP-based techniques (50 µm) [146, 147]. For VAT techniques, fabrication speed is among the fastest in 3D printing, reaching 10^6 mm^3/h or ten times more for DLP and CLIP techniques [146, 149].

Materials selection for SLA is very limited, as they all need to be photocurable resins. This limit is not so visible for construction elements fabrication or pure demonstrators, while within photocurable resins we can find many different materials with various physical properties, especially with tailored mechanical properties or a wide range of colors. The viscosity of photopolymers also should be low (<5 Pa·s), which limits the application of highly loaded composites [141, 157]. At the same time, the fabrication of electrically conductive or bioactive materials is very difficult or sometimes not possible due to the chemistry of the photocurable resins, especially the use of UV activated photoinitiators [158]. More and more new composite materials are available on the market and are presented in the literature [159].

Finally, the VAT techniques have many advantages related to a very smooth surface in comparison to FDM or powder-bed techniques, exhibit higher feature resolution and high production throughput and achievable with relatively low-cost equipment [149,

160]. At the same time, unfortunately, SLA and similar techniques combine the two most important disadvantages of additive manufacturing that limit the applications for 3D printed electronics. First is the limitation to a single material printing with very few perspectives for a breakthrough in this area (some perspectives will be presented in later sections), and additionally, materials based only on photocurable resin that limits the fabrication of electrically conductive components.

2.6 Powder bed techniques

Besides the VAT polymerization techniques, another very important player in the additive manufacturing market is the group of powder-based techniques, also located in the regions of selective material curing or binding, instead of selective deposition like FDM. Today, their application in the fabrication of 3D printed electronic elements is very limited. They have a big potential, which is already confirmed by their vast application in the automotive and aerospace industry, or in biomedical applications for implant fabrication. Contrary to VAT, powder-based technologies utilize a material in a solid state (powder). There is almost no limit on the material type, as they can be from any group of functional materials such as polymers, ceramic, metal or even more exotic like sand or sugar. Powder bed techniques share many similarities to the previously mentioned VAT polymerization techniques, and in some solutions, the modification is as simple as substituting the resin with powder and applying a more powerful laser for sintering instead of curing. Besides the laser, there are other techniques for selective solidification of powders, such as electron beam or printed binder. All powder-based techniques share the basic principle that each deposited powder layer is converted into a continuous solid pattern, achieved by Selective Laser Sintering (SLS), Selective Laser Melting (SLM), Direct Metal Laser Sintering (DMLS), Electron Beam Melting (EBM) or by printing a binder on the powder (Binder-Jet), occasionally sintered with IR lamp in the Multijet Fusion technique (MJF) [161–166]. Sometimes these techniques require separate postprocessing in furnaces for sintering bind metal powders such as the Metal-Jet technique [167, 168]. The general description, an example of an SLS technique, is that a layer of polymer, ceramic or metal powder on a bed/table is sintered with a laser [165, 166], and after that the bed retracts; a fresh powder layer is deposited with the rolled, and the process repeats. For some techniques (SLS, MJF), the powder usually supports the 3D construct, and no additional supporting structures are needed—they are required mostly for FDM, SLA or Multijet techniques. Here, resolution limitations are both related to the size of the powder particles and the spot of the energy beam or ink drop, achieving a feature size of 50–100 μm [165, 166]. Similarly as for FDM, SLA or inkjet the XY and Z resolution is also kinematics-related (movement of the build plate, scanning resolution and focusing of the optics, or resolution of the printing pattern from nozzles). On one hand, one of the biggest advantages of the powder bed group is the fairly large availability of materials, also electrically conductive metals or other functional materials in the form of

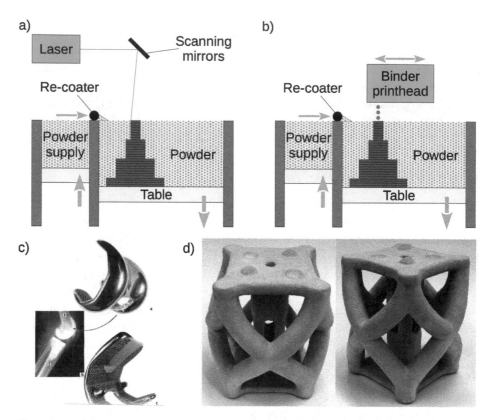

Figure 2.7: (a) Schematic illustration of Selective Laser Sintering (SLS) process. (b) Schematic illustration of the Binder-Jet process. (c) Knee inserts from a Co–Cr–Mo alloy fabricated with the EBM technique [161]. Reprinted with permission from Elsevier, copyright 2017. (d) Example of Binder-Jet 3D printed element from hydrated cement [169]. Reprinted with permission from Elsevier, copyright 2019.

powders. Mechanical properties are comparable to or even outperforming milling or molding techniques (SLM, DMLS, EBM). Process-related advantages cover supportless printing of the structures or no thermal stresses for the Binder-Jet technique. On the other hand, some disadvantages are related to the high porosity of the printed parts, thermal stresses have to be considered for high-temperature laser or electron beam processes, causing distortion to the printed part [162, 166, 169]. Also, similarly as for the VAT polymerization techniques, the drawback for electronics applications is "one powder at a time" approach, thus some attempts for multimaterial fabrication will be presented in the further section of this book. Figure 2.7 shows the working principles of the SLS and Binder-Jet techniques with examples of elements fabricated by these approaches.

2.7 Other techniques

The additive manufacturing family covers a whole lot more different techniques allowing the fabrication of 3D components. One of the most known techniques is Laminated Object Manufacturing (LOM), where materials come in sheet form, the layers are bonded together by external factors such as adhesive, pressure or heat, and the pattern is cut with a laser or a knife [170–172]. The main materials used here are paper of polymer foils, thus metal or ceramic foils are also possible. LOM is an example of the technique proving that the Z resolution, related to the sheet thickness, should not be taken as the ultimate factor determining resolution, although this does not affect the details of the elements directly. Applications of LOM for electronics are very limited, thus in a one-sheet format with printed conductive elements this would be the pure example of printed electronics. Other emerging additive techniques worth mentioning are concrete extrusion for house building [173, 174], sharing many similarities with DIW techniques, and only the adaptation size is much bigger, Big-Area Additive Manufacturing (BAAM) being sort of very large scale FDM [175–177] and Wire-Arc Additive Manufacturing (WAAM) incorporating wire-arc welding and a 6-axis robotic manipulator [178, 179] being a sort of very large scale FDM incorporating metal for fabrication instead of polymers. Direct Energy Deposition (DED) involves the melting of metal powder with a high-powered laser beam or plasma arc and ejecting it at high speed onto a build plate [180], which sounds very similar to the AJP technique, again on the larger scale and with the implementation of high temperature and molten metals. Finally, an adaptation of liquid and molten metal inkjet, FDM and DIW techniques [73, 181–184] or hybrid techniques such as Composite Based Additive Manufacturing (CBAM) covering the LOM, powder bed, sintering, inkjet and subtractive techniques [185] only proves that innovations in additive manufacturing are almost limitless, and we should expect more to come in the next years.

3 Printed electronics

The term "printed electronics" refers to functional electronic, optoelectronic or mechatronics structures and circuits produced using printing techniques [186, 187]. It combines economical mass production methods known from the printing industry with the application of a new class of functional materials including nanomaterials and conductive polymers. Printing techniques are also a part of the additive approach to manufacturing, due to the nature of material application on the substrate. With this method, it is possible to make almost any passive or active electronic components such as resistors, capacitors, inductors or diodes and transistors. With the use of dedicated materials in the form of pastes and inks, it allows components to be made on flexible substrates applied from rolls such as films, paper or fabrics. Printed electronics is a new technological field, but with its capabilities offering a wide range of applications, it will play an increasingly important role in the manufacture of electronic circuits. With a few decades of development, it has acquired great interest in research and practical applications and is successfully fueling the growth in materials science and technology. The development of printed electronics in recent years has been inspired by the announcement that circuits can be produced at a much lower cost than with traditional techniques on a much larger scale with much higher production yields. This makes it possible to produce electronic components that can be lightweight, flexible, thin, optically transparent or disposable. It provides a technological advantage for large-area elements in cost efficiency and eco-friendly production. While printing techniques have been known for centuries, only in the last decades have the advancements in printed electronics technologies exploded, which can be attributed to the progressing printing capabilities of the equipment and the development of new materials and functional inks. Although printed electronics and printed circuit boards (PCBs) both share the "printed" term in their names, the second one has very little in common with printing techniques.

An analysis of the global electronics manufacturing market to date shows that printed electronics is becoming one of the more important electronic technologies and is the subject of intensive research in many research centers around the world [121, 186, 187]. Various forecasts and analyses suggested that the global printed electronics market is expected to grow at a compound annual growth rate (CAGR) of 13.6 %, and reached $26.6 billion US dollars in 2022 [188]. This fast growth is fueled by the interest from consumer goods industries, healthcare, logistics, telecommunications, electronics, multimedia or energy. The possibility to replace some of the classical electronic components with cheaper equivalents and flexible solutions with higher mechanical resilience allows new paths for creating completely new products, increasing the functionality of existing electronic solutions and streamlining production processes. However, the main purpose of using printed electronics is not to completely replace current printed circuit technology or passive and active components, including silicon semiconductor structures, but to provide an alternative solution, complementary to current technologies.

https://doi.org/10.1515/9783110793604-003

The performance of currently obtained printed circuits is lower than that of semiconductor structures. In integrated circuits operating at high frequencies, silicon semiconductor chips will still play an important role in high-level electronics solutions where large-scale integration and high performance are required. Printed electronics technology can provide an alternative for the manufacture of low- and high-volume flexible circuits, prototype series, single-use electronics (e. g., biochemical sensors), smart clothing or low-volume production without the need for tooling facilities. The technology is ideally suited for use in the production of circuits for which performance is a secondary parameter, and where rapid deployment time and low manufacturing cost, whether for single-unit or high-volume production, play a significant role [189].

As the name suggests printed electronics, in short words, is an application of printing technologies to fabricate electronic devices using specially adapted functional inks and pastes. Therefore, it is based on the printing techniques well established in the printing industry used for the production of books, newspapers, packaging, graphics etc. For the production of printed electronics' components and circuits, many printing techniques are used and tailored such as screen printing [190–193] and inkjet printing [194–198], which belong to the sheet-fed printing group, as well as offset [199, 200] flexography [201, 202] and rotogravure [203–205] printing, which belongs to the roll-to-roll printing group. The other division into two categories covers conventional printing techniques and the already mentioned digital printing techniques [206, 207]. Conventional techniques require a physical printing mask, plate or stencil to transfer material to a substrate with the use of the mask and forced mechanical contact. These techniques also require the fabrication of patterned masks, often introducing high setup costs and long preparation times. Typical examples are screen printing, gravure and offset. Digital printing, already covered by the description of inkjet or AJP [126, 133, 134, 208] technique fabricates the pattern from digital data by dispensing/jetting material droplets without direct contact of the nozzle with the substrate. While printing directly means producing patterns by transferring material onto the surface of a substrate, either with mechanical force, hydrodynamic tension, electric field, radiation, etc. many different technical solutions has been developed for decades or centuries for the production of text or graphic [209]. But despite its long traditions, printing has not been easily transferred to the production of electronics, due to the higher printing quality requirements. Graphical images can be constructed from dots forming the desired pattern and do not necessarily need to be a line or film to be interpreted by the human eye as a continuous pattern. Unfortunately, this is not the case for electronics, where strict physical contact must be ensured for the current flow, and even unnoticeable by the naked eye, defects in printouts can cause open-circuit. This requires tailoring printing parameters and the development of new materials that would provide higher printing quality and electrical properties at the same time. It is worth mentioning that while the term printed electronics often also covers such as spin, spray, dip, slot die, blade and bar coating, they are used for processing transparent electrodes or large-scale photovoltaics, easily scaled up and adaptable to large-scale roll-to-roll production. But in this book, we will focus only on the digital printing techniques for selective patterning of devices and circuits, which is important due to the adaptation of selected 2D printing techniques for the 3D printing of electronics.

Having a printing technique, the second element for the printed electronics technology is a material in the form of paste or ink, usually as a composite mixture of an organic vehicle and a functional phase. Conductive polymer solutions can be alternative and they do not contain powder fillers, as the whole volume of the vehicle also acts as the functional phase. The dominating form of materials used in printing techniques, also for printed electronics, are liquids with various viscosity and other rheological properties. The liquid form enables efficient printing or transferring the material on the substrate [207, 209]. Different sets of layers fabricated from functional materials are printed and overlap each other to fabricate complex printed electronic devices. These functional materials can be organic or inorganic, working as conductors, insulators, dielectric or semiconductors or other functional elements (light emission, sensing, etc.) [85, 210, 211]. Using suitable material combinations and component designs, printed electronics can be even deformable, which helps to maintain comfortability and inconspicuousness for future visions of wearable electronics and human body embedded devices [80, 212–214]. It is fair to say that almost all electronic components require conductive electrodes. Here, the range of available materials is the broadest [215–217]. It all started on a large scale with screen printing technology used for decades to produce conductive layers with metallic fillers, resistors with carbon fillers and also dielectric layers [218–220]. Now the majority of conductive materials are metal nanoparticles (NPs) and nanowires (NWs) synthesized from silver, copper, gold or other metals [221–228], carbon nanotubes (CNTs) [229–235] and graphene platelets (GNPs) [99, 236–240] but also conductive polymers [241–243], nanoinks with transparent conducting oxides (TCOs) [244–246], semiconductor nanomaterials [195, 247–250] and recently reactive inks metallic compounds and complexes as precursors (nanomaterials-free) converted to bulk metal after deposition [251–254]. Photonic devices such as displays or organic light emitting diodes (OLEDs) and photovoltaics (OPVs) require transparent electrodes [255–257] and semiconductors as nanomaterials dispersion or conjugated polymers solutions (also used for printing transistors) [258–260] not to mention eco-friendly materials for disposable electronics [261–264] or patchable [265, 266] and ingestible bioelectronics [30, 267]. In most cases, the final functionality of the electronic product is related to the physical properties of the ink (conductivity, optical transparency, mechanical stability to bending, twisting and stretching, etc.) of the ink.

Printed electronics applications vary from simple paths, electrodes or antennas requiring conductive layers [268, 269] up to complex transistors and integrated circuits [134, 270], energy devices [271–274] or wearables [275–277]. One of the most common applications is Radio Frequency Identification (RFID) electronic tags printed from metallic pastes [278, 279]. There is also a strong emphasis on the production of printed electronic circuits for the manufacture of photovoltaic cells. Initially, screen printing was used to make conductive paths on silicon substrates, because as the only contact printing technique, it allows printing without exerting pressure on the substrate, thus reducing the number of damaged silicon substrates [280]. But screen-printing is commonly used to print electroluminescent structures using inorganic ZnS:Cu powders on a mass scale [281, 282], electroluminescent displays with OLED polymer diodes [255, 283, 284], temperature and pressure sensors [285, 286] and various sensors for chemical and biological analysis [287, 288]. Printing also finds application in the fabrication of photovoltaics of photoactive conductive polymers [289–291] in photoelectrochemical cells called Grätzel cells [292] and Cu-In-Ga-Se (CIGS) layers characterized by one of the highest energy yields among photovoltaic materials [293, 294]. There are also attempts to make printed organic field-effect transistors (OFETs) [295] and even printed Nonvolatile Random Access Memory (NVRAMs) [296] as well as printed energy storage sources such as batteries and capacitors [297, 298]. Examples of electronic components and circuits made using printed electronics technology are presented in Figure 3.1.

The advantages of printed electronics technology include the ability to produce large-area structures and continuously printed circuits, multilayer circuits, the increased mechanical strength of circuits related to the flexibility of materials, the ability to print on any type of substrate, including flexible substrates, as well as environmental aspects such as easy processing or the ability to produce biodegradable and disposable circuits. Among the most important benefits are low-cost alternatives to conventional electronics components, also for low-volume and high-customization productions [189]. At the same time, this is an efficient way for high-volume fabrication with roll-to-roll (R2R) printing techniques and large-area processing such as solar panels or information displays [121, 186, 299]. A wide range of substrates can be used, rigid and flexible. And finally, lesser material wastage promotes a more environmentally friendly approach to electronics technology [300]. Some disadvantages also emerge, such as low printing resolution compared with conventional electronics technology (PCBs or ICs), with relatively high surface roughness and thickness variations affecting performance [301]. And in the light of additively manufactured structural electronics applications contact printing techniques with masks or stencils severely limit or even make impossible the adaptation for 3D printing. Therefore, only maskless, noncontact, digital printing techniques (inkjet, AJP) are attracting significant attention in the area of additive manufacturing, enabling translation from 2D to 3D printing.

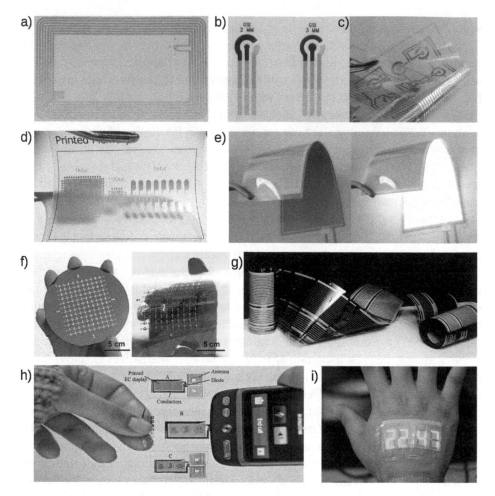

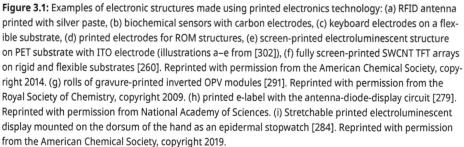

Figure 3.1: Examples of electronic structures made using printed electronics technology: (a) RFID antenna printed with silver paste, (b) biochemical sensors with carbon electrodes, (c) keyboard electrodes on a flexible substrate, (d) printed electrodes for ROM structures, (e) screen-printed electroluminescent structure on PET substrate with ITO electrode (illustrations a–e from [302]), (f) fully screen-printed SWCNT TFT arrays on rigid and flexible substrates [260]. Reprinted with permission from the American Chemical Society, copyright 2014. (g) rolls of gravure-printed inverted OPV modules [291]. Reprinted with permission from the Royal Society of Chemistry, copyright 2009. (h) printed e-label with the antenna-diode-display circuit [279]. Reprinted with permission from National Academy of Sciences. (i) Stretchable printed electroluminescent display mounted on the dorsum of the hand as an epidermal stopwatch [284]. Reprinted with permission from the American Chemical Society, copyright 2019.

4 Structural electronics

The history of electronic development can be summarized with one word—integration. Almost all of the inventions implemented in the electronics technology production chain were focused on integrating multiple functionalities within smaller components and devices as possible. The goal is to integrate into a single utility more components and more functions, make the construction simpler and more efficient and at the same time not compromise other aspects of the final application. Embedding electronic functionality within an enclosed part offers many benefits: fewer parts and connections, minimal assembly, lower weight, positive impact on sustainability and simpler supply chains of components. One of the ultimate examples of the constant struggle to integrate more and more functions into one device is a smartphone. Internally, it is based on a "regular" computer for operating various information tools, such as messengers, maps, games, cameras, notes, calendars and thankfully still phones, hence the name. In fact, besides faster and more powerful computers, smartphones are also equipped with a whole set of additional components, such as camera sets, sensors, navigation systems, additional communication (BT, WiFi), chargers and battery, sometimes stereo audio, very large display, ports for memory cards, headphones or USB and lots more. A few decades ago, the phone was set on a table with a dialing pad, handset and a casing. Now smartphones are more powerful and are able to perform more operations than NASA computers in the command center operating Apollo missions [303]. All that is possible thanks to the integration of components in one utility, and that is possible thanks to the miniaturization and progress in the packaging of electronics.

But still, this is packaging from single components or dedicated integrated systems on one "mainboard" mounted in a frame, chassis, casing or any other form of additional construction component. The next paradigm shift for electronics is the introduction of new forms, which drives the invention of new manufacturing approaches that rely on new materials. All of that will be covered in further sections of this book. Some of these technological advances are already in our hands, but there is a need for new concepts of electronics for our daily life applications or for specialized work and missions. And here the idea of a new form of electronics arises. Let us ask a question: what if we skip the assembling step in the electronics technology production chain, eliminating costly procedures of PCB fabrication and packaging of hundreds of components, and organize the production process in a way that the desired functionality of the device will be integrated into the object itself? This is the world with "structural electronics," where the boundaries between mechanical components and electrical systems overlap or even blur and daily life components have all integrated electronic functionality inside them.

Structural electronics is a term describing electrical and electronic components and circuits that act as construction elements, casings or other types of protective dumb structures (like vehicle bodies), embedded inside the volume of element or conformally placed on the surface [304, 305]. This approach reduces weight and complexity in manufacturing allowing for more versatile design options eliminating rigid PCBs if they are

https://doi.org/10.1515/9783110793604-004

not needed. Due to the possibility of enhancing product functionality, lowering production cost or providing the possibility for fabricating consumer-tailored devices, structural electronics is an interesting new branch of technology for aerospace and military applications, home appliances, consumer electronics, automotive industry and civil engineering. The realization format for structural electronics is mostly provided by fitting standard geometry components into spare spaces in machines, tailoring the shape of conventional components to fit into spare space, and adding conformal coating—a "smart skin"—on the surface of existing elements (sensors, photovoltaics), in-mold shaping of multilayer electronics (home appliance casings or car audio) or finally forming as load bearing elements (bridges, houses, cars). Enhancing functionality by the addition of "smart skin" or elastic-shaped elements can be realized by using printed and elastic electronics, but the load-bearing or construction elements need more complex solutions, like laminating, in-mold shaping, injection molding, casting or finally 3D printing (Figure 4.1(a)). This is a complete paradigm shift for the design and production of not only electronics but most surrounding objects, line furniture or houses, and even today this grand vision offering high potential for innovations still needs a lot of research and development work in the areas of new fabrication techniques and new materials, which is covered in this book.

Before going into the details, let us look back at the history of such implementations. The origins of structured electronics can date back to the late 1990s, when the US Defense Advanced Research Projects Agency (DARPA) announced the Solid Freeform Fabrication program, and a few years later started the Mesoscopic Integrated Conformal Electronics (MICE) program, aiming to produce electronics, sensors or antennas on nonflat surfaces such as helmets or other equipment [306]. During this project, direct write platforms were developed that allowed electronic materials to be applied to complex geometries using printing techniques such as dispenser printing and aerosol printing [307]. An auxiliary goal behind the MICE project was to answer the call of another developing concept at that time, the Internet of Things (IoT). This concept involves adding new functionalities to common objects for collecting, processing and transferring data between each other over a network. This will not just allow greater convenience in everyday life, but also better management of industrial production, cities or the further development of medicine [308, 309]. The main guidelines that structured electronic circuits should meet were established:

- integration of standard electronic components;
- application of polymers both as the substrate of the electronic circuit and the housing of the device;
- employing manufacturing processes not exceeding 100 °C, allowing the use of common polymers;
- designing circuit layout in 3D for better use of device space;

A very important aspect of the MICE program was the development of a technology that would enable a single process to produce a finished circuit in a matter of hours, based on

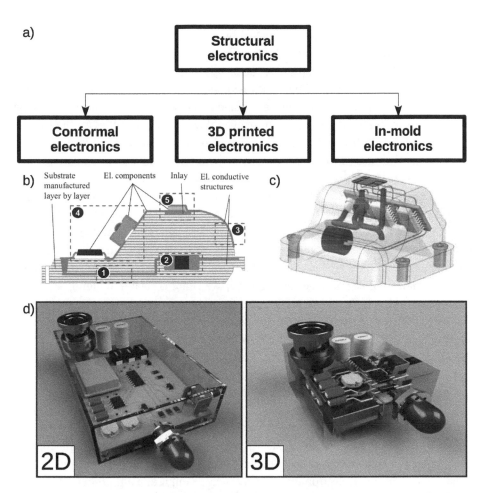

Figure 4.1: (a) Schematic categorization of structural electronics, divided into the main groups related to the fabrication technology [302]. (b) [310], (c) [311] and (d) [312]. Opportunities for the adaptation of structural electronic circuits with embedded conductive tracks and components. (b) and (c) Reprinted with permission from Elsevier, copyright 2014. (d) Reprinted from MDPI, Basel, Switzerland, under Creative Commons Attribution 4.0 license, copyright 2022.

a tailored design—an approach taken from rapid prototyping techniques. At that time, transferring this idea to electronic circuits was challenging because of the required variety of materials and the combined approach of fabricating the circuit, mounting the electronic components, making connections and building the casing at the same time. The program also assumed that the fabrication systems will be able to apply different types of materials for fabricating both passive components (resistors, capacitors, antennas, connectors) and active components (photovoltaic and fuel cells, batteries, sensors). With full automation of the procedure and the elimination of additional soldering, it was expected that structurally developed devices would exhibit higher reliability [306].

Moreover, such an attempt with the development of low-temperature curable materials will allow the production of electronic circuits on virtually any type of substrate: polymers, metals, glass, ceramics, paper, etc. The program provided many valuable concepts and spin-off companies that will be presented in detail in further sections. Most of all, it started a completely new path for the development of new forms of electronics and new functionality for utilities surrounding us in everyday life. Opportunities for the adaptation of structural electronic circuits with embedded conductive tracks and components are presented in Figure 4.1(b, c, d).

But structural electronics is not only an additive approach for the fabrication of electronics as in the MICE program. Generally, this is the concept of the ultimate integration of electronics with all other surrounding us "stuff." The primary needs identifying the implementation of structural electronics are emerging from the application. Adding new functions to the component is usually the primary need in most of the applications. However, other attributes are not always versatile. While saving weight is not very crucial in civil engineering, and saving space is not crucial in buildings or boats, these factors will play a major role in the aerospace and automotive industry or mobile electronics. At the present time, we should ask the question: which components are most suitable for fabrication and implementation in structural electronics? Simple conductive elements, antennas, interconnects, actuators and photovoltaics can be readily adapted in the volume/surface of the construction elements, moving parts, narrow spaces or on curved shapes. A bigger problem is with power elements like supercapacitors, batteries, fuel cells and electric motors. These elements cannot be fabricated as surface coatings (with the exception of capacitors and batteries), and usually they will carve large volumes of elements. Answering this question will direct us to the application of structural electronics, or even defying which elements in our surroundings are already with embedded electronics inside. A catchy example and not so obvious of structural electronics in practice is a new Saint Anthony Falls Bridge in Minnesota, US, called America's smartest bridge [313]. The previous one collapsed in 2007, so the new one is stuffed with sensors monitoring its condition. In the automotive industry, there are two adequate examples from Ford and Volvo. Ford Fusion overhead control cluster is fabricated with in-mold electronics, saving up to 40 % weight/space/cost and increasing reliability [314]. The project realized by Volvo and Imperial College London is to fabricate supercapacitors in the form of car bodywork to save space in electric vehicles [315].

The development of structural electronics has given rise to a number of techniques for its manufacturing. The most common division is based on the geometry of the electronic circuit (mostly conductive paths) and its relation to the substrate, and distinguishes between conformal electronics, with paths and components on the surface of the device, and fully three-dimensional structural electronics, where conductive paths and components can be hidden inside the load-bearing components or casing of the device. The first approach requires manufacturing separate substrates for the electronics, when in fully three-dimensional techniques it is possible to make virtually finished

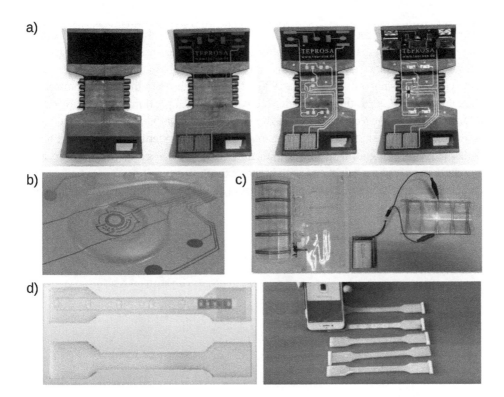

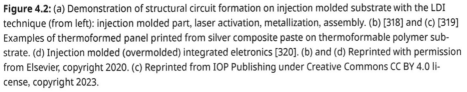

Figure 4.2: (a) Demonstration of structural circuit formation on injection molded substrate with the LDI technique (from left): injection molded part, laser activation, metallization, assembly. (b) [318] and (c) [319] Examples of thermoformed panel printed from silver composite paste on thermoformable polymer substrate. (d) Injection molded (overmolded) integrated eletronics [320]. (b) and (d) Reprinted with permission from Elsevier, copyright 2020. (c) Reprinted from IOP Publishing under Creative Commons CC BY 4.0 license, copyright 2023.

devices with a single process. An example of a conformal approach is Laser Direct Structuring (LDS), used for years for integrating electrical features onto 3D surfaces [316], presented in Figure 4.2(a). By activating the surface of the injection-molded plastic part in a pattern with a laser and subsequent metallization using electroless plating, this enables the creation of hundreds of millions of devices annually, with antennas being the most distinctive example. A very good alternative to conformal and 3D forming is an intermediate technique called In-Mold Electronics (IME) [317–320] employing the two-stage manufacturing process: forming electronic system on a flat, thermoformable substrate with printing techniques and attaching SMD components using conductive adhesives followed by plastic molding of the final embedded system (vacuum forming or injection molding) when the device receives its final shape (Figure 4.2(b–d)).

Structural electronics have immense advantages of delivering new functionalities, using less material and less assembly reducing costs, but it also presents many technical difficulties. The most distinctive one is creating conductive and functional materials tailored for dedicated techniques (injection molding with high temperatures, composite processing with high load of function fillers, etc.). The same applies to the integration of discrete components, such as resistors or ICs [310–312]. Methods for efficient and reliable connections need to be addressed, and that also relates to the use of specialized materials. For IME also, the overall production yield is important, while both purely decorative quality must be extremely high as well as reliable fabrication of electronics circuit—if there is a single problem with one of the process steps, the entire part must be discarded. In the end, it turns out that integration provides benefits, but also creates challenges, and we might end with higher expectations of components being both appealing for decorative purposes and not just providing efficient functionality of electronics themselves.

5 Materials for printed and structural electronics

The crucial aspect fueling the development of structural electronics is functional materials. Besides the new and adapted additive techniques progress in functional materials is one of the biggest drivers, opening new possibilities for the integration of functional electronic components in one printing set. This was most visible in the development of printed electronics, where printing techniques were ready to deploy, but only the new electronic inks and pastes were the starters for the revolution. Here, we are facing the same situation. This is why materials are in the spotlight of scientific and industrial centers around the world.

Regardless of the technique, conventional or additive, electronic circuits require a range of materials with suitable properties, primarily conductive and insulating, also with other functions (i. e., semiconductors). In PCB, fabrication as conductive materials, metals with high conductivity, such as copper, silver, nickel and tin, are most commonly used. Insulators are usually substrates, very often made as laminates of paper or glass fiber layers, bonded with epoxy or phenolic resins (e. g., FR4, FR2). Other functional materials with properties such as magnetic, electroluminescent or semiconducting are also used in electronics manufacturing, mostly for individual electronic components. The same applies to printed electronics technology, where electrically conductive materials are usually polymer composites with an active phase in the form of metal powders (silver, nickel or copper) and graphite fibers and flakes. Thick-film polymer resistors are made of carbon black or graphite-filled composites containing fine carbon black or graphite powders, and dielectric materials are composites with ceramic powders (i. e., $BaTiO_3$). Other materials used to manufacture specialized structures include thermally conductive or thermal insulating pastes, photovoltaic or luminophore emissive layers and others such as solid-state electrolytes. There is no reason why it should be otherwise for structural electronics, especially 3D printed. With the use of FDM, inkjet, AJP or DIW, conductive, insulating and functional inks, pastes and composites will also be used to form the final solution [321–325]. In addition to the requirements for their functional electronic properties (conductors, dielectrics, semiconductors, etc.), it is necessary to process them in the desirable form, like liquid solutions or suspensions, with a viscosity adapted to the printing technique, for materials used in printed electronics. The type of material (organic, inorganic) is a secondary consideration—typically inorganic materials are in the form of suspensions (heterophase compositions) and organic materials in the form of solutions in solvents. Examples of suspensions are metallic inks and pastes with micro and nanoparticles of silver, gold, copper, etc. but may contain another type of functional phase, such as carbon fillers. The thing is that in most cases (besides inkjet and AJP techniques) it is difficult to directly transfer materials from printed electronics to structural electronics fabrication without major modifications (like screen-printing pastes for DIW) or developing completely new groups (composites for FDM). Figure 5.1 shows the classification of functional materials adaptable for 3D printed structural electronics.

https://doi.org/10.1515/9783110793604-005

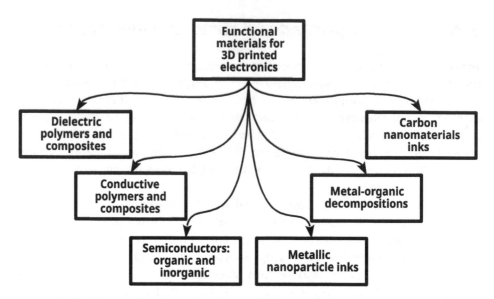

Figure 5.1: Classification of functional materials adaptable for 3D printed structural electronics.

5.1 Conductors

By far, the best-developed additive manufacturing technique favorable for the additive manufacturing of electronics is FDM, having one of the most diverse material portfolios. This technique utilizes easily adaptable thermoplastic polymers and composites, therefore, it was natural that most of the work today is focused on the fabrication of conductive composites. Such materials have been known for more than a century in various forms, but the general working principle and composition remain similar [326, 327]. The conductive inclusions of the functional phase realize the conductivity of electrons, whereas the polymer (most often insulator) acts as a matrix suspending conductive fillers. For techniques such as FDM (can be adapted as composite powders for SLS), commonly used conductive fillers are metallic powders and various carbon forms such as carbon black, graphite, nanotubes or graphene. The most widely used and easily adapted materials are composites with different sorts of carbon, including carbon nanomaterials [99, 328–330]. The work is progressing on the application of copper-based composites achieving much lower resistivity than carbon materials [66]. Because highly reactive copper oxidizes rapidly in contact with air, forming dielectric oxides, other materials are considered such as nickel or iron/steel particles [331, 332]. The most popular conductor used in printed electronics, silver powders in the form of inks and pastes, is too expensive to be utilized for highly loaded polymer composites. This is why it is not explored in research and not applied into the market. The same applies to gold, which is even more expensive than silver. One of the most popular conductors, aluminium, an abundant and cheap material, is one of the most problematic ones for the application in conductive composites, as it exhibits similar reactive properties as copper, and Al_2O_3

is one of the best dielectrics used in electronics today. To provide electrical conductivity in an isolation polymer, a minimum required concentration of conductive fillers needs to be met, known as the percolation threshold, described in detail later in this chapter [333]. Conductive composites are relatively cheap and offer a wider variety of material options, yet their electrical conductivity is moderate and highly dependent on the particle type, morphology and concentration, also affecting printing capabilities. New composite materials branded as conductive filaments, while gaining more attention, do not present satisfactory properties for consumer electronics, not to mention special applications. The primary obstacle is high resistivity, but in many cases lowering resistivity will not solve all the problems, to mention only a higher current density, lower noises and higher reliability that is required for high-efficiency electronics.

Besides the solid metals and composites, the other large group are composite pastes and inks in liquid form. Most metals are solids at room temperature, but printing techniques require the material to be in liquid form for efficient application, and the easiest way to achieve this is to prepare a dispersion. For decades, electronic pastes were used in thick film technology to make electrodes, resistors and other components. Initially, such pastes were used in high-temperature processes (containing glass, sintered layers over 800 °C), or in the form of polymer pastes or adhesives for heat-sensitive substrates, cured after application by evaporating solvent or cross-linking of the polymer [218, 334]. The advent of nanotechnology has led to the development of a material combining the advantages of both types of paste mentioned above. Materials with metal nanopowders are sintered at relatively low temperatures around 250–300 °C, which led to the development of pastes with specific conductivities reaching more than 60 % of that of pure silver, which can be used on flexible polymer substrates [335, 336]. Alternatively, instead of metals, carbon based pastes are also available. Initially, carbon black or graphite materials were used as low-cost conductive coatings or simple heaters due to their low cost and sufficient conductivity. With the discovery of carbon nanomaterials, more focus has been set on these materials, that besides very high electrical conductivity, they exhibit high optical transparency, high flexibility and mechanical strength and good chemical stability [285, 286, 337–339]. All these materials in the form of viscous pastes used for screen-printing and other sheet techniques are adaptable for the application to 3D printing with a DIW technique. As mentioned before, DIW requires that the paste maintain its structure after extrusion from the nozzle in the ambient air, optionally assisted with external factors. Therefore, in various studies on DIW most of the materials come in the form of viscoelastic dispersions, colloidal pastes of metallic powder into a polymer binder or organic solvent with adequate viscosity [85, 86]. While the most popular form of functional conductive filler is powders, conductive printing materials can also deliver metals in other forms. The source of the metals is often in the form of salts or other chemical compounds, which is why such materials are often called organometallic inks, precursor-type inks or nanomaterials-free inks, and in general metal-organic decomposition (MOD) inks [340, 341]. After the deposition, a thermal sintering process at around 200 °C is performed for the decomposition of complexes into conductive metals and waste gases, resulting in a conductivity of 10^6–10^7 S/m. These inks can contain a wide range of metals. There are materials containing aluminium [342, 343] or nickel [344], but the best electrical properties are achieved by inks with silver [345, 346] gold [347, 348], platinum [341] and the most promising copper [349–352]. The MOD materials exhibit low cost, and they are several orders of magnitude cheaper than metal nanopowders, which favors them as the materials for low-cost additive manufacturing.

Considering the viscosity properties of the materials, the last remaining group of printing techniques will utilize low-viscosity inks, as it is for inkjet and AJP techniques. Generally, the major difference is in the composition focused on the rheology of the ink, and thus lower viscosity than in the above mentioned pastes. The composite inks usually are based on similar functional materials, including metal nanopowders, carbon nanomaterials or MODs [130, 211, 353, 354]. The use of low-viscosity inks is in some way limiting the use in electronic applications, while highly conductive materials requiring high loading of functional phase, even based on nanoparticles could lead to nozzle clogging [120]. At the same time, MOD inks are particle-free and have longer shelf-life compared to nanoparticle inks. So, nozzle clogging is less likely to occur.

5.2 Semiconductors

Semiconductor materials are used for the fabrication of active electrical components such as diodes and transistors (field-effect transistors (FETs); thin film transistors (TFTs)), but also for sensing purposes or energy conversion in photovoltaics. Printed active components require inks' semiconductor properties such as bandgap, on/off ratio and mobility. The two main types of semiconductor inks are inorganic dispersions with metal oxides or nanomaterials, and organic-based solutions or dispersions with conjugated polymers.

The first group, inorganic dispersions, shares many similarities with conductive composites, pastes and inks. Instead of metallic powders or nanoparticles, here most often metal-oxides, semimetals or compounds (i. e., III–V or II–VI) are adapted [355, 356], but at the same time they are almost identical to previously described group in terms of carbon nanomaterials application [357, 358]. The new group of materials here are also quantum dots or MXenes that introduce new possibilities of application for electronics and photonic devices [359, 360]. Such materials exhibit fairly high environmental stability and electronic transport properties, but often require thermal treatment after deposition. Hence, not all of them will be suitable for the fabrication of active electrical devices on temperature-sensitive substrates. The most popular inorganic oxide semiconductors are from the group of indium oxide (In_2O_3), indium–tin–oxide (ITO), indium-zinc-oxide (IZO), indium–gallium–zinc–oxide (IGZO) and zinc-tin-oxide (ZTO), but also tin monoxide (SnO), cupric (CuO) and cuprous oxide (Cu_2O) and zinc oxide (ZnO)

[356]. Besides inorganic metal oxides, which can be in the form of micro or nanopowders, a wide group of nanomaterials such as carbon nanotubes, graphene and other 2D semiconducting materials including tungsten disulfide (WS2), tungsten diselenide (WSe$_2$) and molybdenum disulfide (MoS$_2$) may be used [361]. The various characteristic features, commonly used materials and applications for each type of functional material are tabulated in Table 2.

Alternatively, a group of conductive polymers can be used for the preparation of active components such as organic light emitting diodes (OLEDs) [283], organic field-effect transistor (OFET) or organic thin-film transistor (OTFT) [362–364] organic photovoltaics (OPVs)[291, 365, 366] and sensors [367, 368]. The most common mechanism of conductivity in these polymers is the occurrence of delocalized electrons in a π-conjugated backbone structure in the main carbon chain [369, 370]. The most commonly used compounds are polyethylene dioxythiophene (PEDOT), polyacetylene (PA), polyphenylacetylene (PPA), polyphenylene sulphide (PPS), polyfluorene (PF), polyaniline (PANI), polypyrrole (PPy), polyphenylene vinylene (PPV) and poly(3-hexaylthiophene) (P3HT). These substances are generally characterized by good mechanical stability and adhesion to polymeric substrates, flexibility and low weight, but at the same time for most of them, electrical conductivity degrades in contact with air and humidity or in elevated temperatures (even at 150 °C). Only a few of the conductive polymers are easy to dissolve (e. g., PEDOT:PSS copolymer) [371, 372].

Here, it needs to be clearly stated that the majority of the materials described in the semiconductors section are suitable for the deposition with ink-based techniques (inkjet or AJP) or at least with the DIW technique, due to the nature of the structures fabricated with them—thin or thick film transistors and other components. There are no representative reports in the state of the art about incorporating solid composites with semiconducting properties into the FDM technique.

5.3 Dielectrics

The last main group of materials are dielectrics and insulators, with the primary objective to exhibit electrically insulative properties. The simplest approach is to apply polymer, the most common material for additive manufacturing, as a "non-functional" phase, while in their nature they exhibit dielectric properties suitable for most of the electronic applications for multilayer insulation of circuits or dielectric layers in capacitors and transistors [373]. Besides the regular form of organic polymers, dielectrics can be in the form of inorganic composites for FDM or micro- and nanomaterials suspensions for ink printing. The most common organic polymers for dielectrics include acrylonitrile butadiene styrene (ABS), polylactic acid (PLA), polycarbonate (PC) polymethyl methacrylate (PMMA), polyimide (PI), polystyrene (PS), polyvinylidene fluoride (PVDF), polyvinylpyrrolidone (PVP) or polyvinyl alcohol (PVA), just to name a few. Often, dielectrics should exhibit other properties adjusted to the deposition process, such as low-temperature processability, which can be altered for composites, or high optical transparency for encapsulations, and preferably low cost [186]. Their specific properties should be adjusted to the final applications such as transistors or capacitors, where dielectric permittivity can impact the final performance of printed components. For instance, capacitors require dielectric inks providing high capacitance density through high dielectric permittivity and low loss, and here inorganic composites are usually preferable over organic dielectrics, having higher dielectric permittivity, higher device stability and lower hysteresis [186, 374].

5.4 Dispersion

While the majority of the materials described here, designed for additive manufacturing, are polymer composites the important aspect that should be taken into consideration is the dispersion of functional phase in polymer matrix. By the dispersion, I mean both the structure morphology of the composites, regarding the volumetric distribution characteristics of the functional phase, and also the procedures implemented for the preparation of composites.

The main problem in the effective fabrication of these materials is the selection of the appropriate procedures to obtain a homogeneous dispersion of powders in the polymer matrix. This is highly important due to the various types of composites that are used in additive techniques (solids for FDM, viscous pastes for DIW or liquid inks for inkjet and AJP). Each of these materials requires separate and tailored procedures for the preparation of final mixtures. Moreover, regardless of the state of the composite (solid or liquid) additional consideration is required taking into account the type of functional phase used: micropowders, nanomaterials, flakes, spherical, fibers, etc. This is most visible in applications of nanomaterials, which being in powder form, are tending to formulate strongly bonded agglomerates that are not easy to break [375]. The visualization of the dispersion structure comparing micro and nanoparticles in the matrix is shown in Figure 5.2(a). Usually, for macro and microparticles, it is easy to obtain a homogeneous dispersion, as adequate separation can be ensured. However, the same volumetric amount of nanomaterials poses problems related to the tendency to agglomerate, caused by the presence of electrostatic attraction and van der Waals forces [376]. The structure of the polymer matrix affected by the dispersion of particles changes, especially if network-forming particles with dimensions close to the near-order distance of the polymer chains are introduced. The changes occur at the level of intermolecular interactions of the polymer chains, due to the developed active surface of inclusions, resulting in the formation of a matrix transition phase (Figure 5.2(b) with altered chemical, rheological and mechanical properties [377, 378]. For the fiber fillers, this regime spreads a few micrometers [379, 380], while for nanomaterials, hundreds of nanometers [381] that thus significantly affect homogeneous dispersion. Another issue with homogeneous dispersion of particles, especially nanoparticles, is the fact that, in powder form,

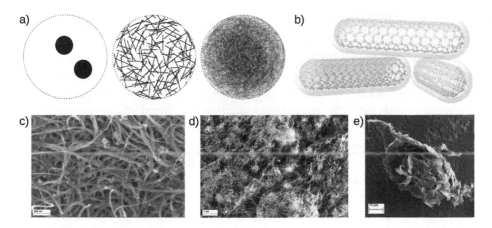

Figure 5.2: (a) Dispersion visualization of macro, micro and nanoparticles in a polymer matrix at the same volume content: (from left) flakes or spherical powders, microfibers, nanotubes. (b) Visualization of interconnections between carbon nanotubes in a polymer composite, with a visible transition phase on the surface of the nanoparticles. Scanning Electron Microscopy micrographs of agglomerated carbon nanoparticles: (c) bundles of double-walled nanotubes, (d) agglomerates of multiwalled nanotubes and (e) agglomerate of graphene flakes. Illustrations a–e from [302].

they tend to form highly bonded agglomerates, demonstrated in Figure 5.2(c–e). Leaving the agglomerated structures in the composite leads to inferior electrical, optical and mechanical properties, as expected from models predicting homogeneous dispersion in the matrix [378, 381, 382]. This suggests that the problem of homogeneous dispersion is not only related to the dimensions of particles and their dispersion in the matrix but also to the breakdown of agglomerates and the suppression of reagglomeration. In the end, the dispersion of particles is crucial not only for effectiveness (low resistivity) but also to avoid poorly dispersed agglomerates acting as large particles that could clog the printer nozzle (regardless of the additive technique used) and jeopardize the printing process.

There are many examples in the literature of techniques used to prepare composites, including nanocomposites. The mixing techniques presented here are not the only ones available, but they are the most widely used, and above all, allow the high yields that are necessary for the production of solid composites, inks and pastes for 3D-printed structural electronics. The two most used methods are a solvent-based approach for creating micro or nanoparticulate dispersion [383] and high-temperature melting and mixing the printing material (e. g., thermoplastics) [384].

Mixing particles in solutions and suspensions is the most common method for producing composites [385–387] perfectly suited for inks and pastes. A typical suspension blending process follows several steps, two of which are always present: dispersion of the materials in a solvent (e. g., ultrasonically) and blending with the polymer matrix as a low- to medium-viscosity liquid (e. g., shear mixing or stirring). Additionally, surfactants may be introduced. Ultrasonic stirring involves delivering mechanical energy with ultrasonic vibrations (in the range of 20÷60 kHz), usually using devices such as an ultrasonic bath or sonotrode. Ultrasonic waves propagate through a medium (e. g., a solvent) and excite vibrations in the particles, also enhanced by the cavitation, causing the separation of particle agglomerates [388]. This is an especially efficient method for mixing and deagglomerating nanoparticles in low-viscosity liquids such as water, alcohols or organic solvents [389–396]. For polymer solutions, significant dilution is required to achieve a low-viscosity solution—this can be increased at a later stage by evaporating excess solvent. However, high mixing energies can lead to the damage of the particle structure [397, 398]. In extreme cases, even graphene flakes and the carbon nanotubes turn into an amorphous carbon [399].

Shear mixing is a process carried out on a machine called a three-roll mill, in which three adjacent rolls rotate at progressively higher speeds, causing the material to homogenize [400–402]. It is commonly used for the medium viscosity materials for the dispersion of pigments in printing inks, varnishes and cosmetics, as well as in thick film material technology in electronics. By controlling the shear rates resulting from the difference in speed of the rollers and the gap between the rollers (in the range of micrometres), we are able to control the degree of dispersion of the material in the matrix. It is possible to repeat the process several times for greater homogeneity, but too many repetitions lead to damage to the particles [400].

Blade, magnetic or propeller mixers are also one of the techniques used for the dispersion of particles in liquid polymer matrices. By controlling the size and shape of the mixer, process speed and time, it is possible to obtain suspensions with a high degree of homogeneity [403]. It has also been observed that better dispersion is obtained for particles with larger characteristic dimensions, but also particles have a much higher tendency to reagglomerate [404, 405].

The most popular techniques used for the preparation of solid composites are ball milling [406–412] or hot mixing with extrusion [413]. Very often these techniques are used after the initial procedures of mixing particles with polymer solutions with one of the above mentioned techniques, and after solvent evaporation the composite is additionally ground to pellets and then mechanically mixed with one of these techniques.

5.5 Percolation

Closely related to the dispersion scope is the effective formation of conductive paths in composites, resulting in their final electrical properties, and thus high or low conductivity. The electrical properties of polymer composites depend primarily on the type of the functional material, i. e., metal or carbon, its structure-like characteristics and dimensions, the mentioned degree of dispersion and finally the type of polymer matrix. One of the fundamental relationships describing the change in electrical properties of composites as a function of the amount of functional phase material is the determination of the percolation threshold, allowing the determination of the critical content of filler material in a conductive composite, causing a change in the properties of the composite from insulating to conductive [404, 414]. The percolation threshold is attained when conductive particles form contacting points throughout the entire polymer matrix. All mentioned parameters (dielectric properties of the matrix, conductive particle dispersion, type and morphology) can affect the percolation threshold. The mathematical basis of the percolation phenomenon was defined in 1957 by Broadbent and Hammersley [415]. A characteristic concept in percolation theory, the percolation threshold p, is a sudden and significant change in the properties of the medium associated with an increase in connectivity. The percolation probability $P(p)$ is defined, as related to the probability that in any area of a given heterogeneous medium, elements are connected to each other enough to initiate flow or conduction phenomena. The model assumes that flow occurs through bonds (e. g., pores, particles) that connect the nearest nodes of an n-dimensional network, either regular or irregular in structure. The nodes of the network are populated with conductive particles in a random manner. The number or volume of particles is denoted by x, and for small values of x the probability $P(p)$ is equally or close to zero. This means that the particles at the nodes of the network are isolated from each other and the whole system exhibits insulator properties. As the number of particles increases, they begin to aggregate and once the critical value of x, denoted as x_c, is exceeded, they form a cluster of infinite size (with respect to the surrounding) and the whole system acquires conductive properties. It can be concluded that an "insulator-conductor" transition is taking place in the system. A schematic diagram of the process of conduction path formation in a two-dimensional network is shown in Figure 5.3. The properties of three-dimensional and n-dimensional systems are determined in an analogous manner.

The general percolation model assumes that percolation probabilities can be represented as follows (5.1):

$$P(p) \sim (x - x_c)^b \tag{5.1}$$

where critical concentration x_c is called the percolation threshold and b is the critical exponent depending on the dimensions of the network. Using the basic percolation equation, Kirkpatric proposed a description of the electrical conductivity of heterogeneous systems [333], consisting of a mixture of an insulating substance filled with conducting particles (5.2):

$$R \sim R_0(V - V_c)^t \tag{5.2}$$

where R_0 is a constant equal to the resistance value of the conducting phase, V_c the critical volume fraction of the conducting phase at the threshold point, t the critical exponent of conductivity for $V > V_c$. The value of V_c depends on the parameters describing the pathway topology, including the distribution, orientation, size and shape of the particles. An optimal set of V_c and t is sought for the relevant experimental data. Concluding,

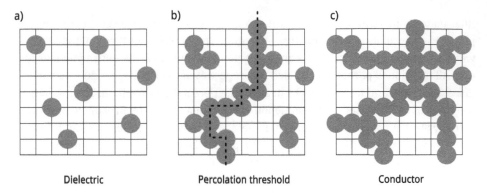

Dielectric · Percolation threshold · Conductor

Figure 5.3: A schematic diagram of the process of conduction path formation in a two-dimensional network.

the nature of the conductivity in composites is not just a linear function of the amount of particles added to the polymer matrix, and many factors influence the final electrical properties. Not to mention, that all these phenomena are related to the feedstock, and later deposition techniques, especially thermal extrusion, can and will additionally introduce external factors further altering the properties of deposited materials, and thus the final 3D printed structure.

6 3D printed structural electronics

Previous sections dedicated to additive manufacturing, printed electronics, structural electronics concepts and innovative electronics materials presented new directions for the development of electronics technology. A visible advantage of additive manufacturing was presented, overcoming many limitations attributed to conventional manufacturing methods like subtractive machining or formative casting in the fabrication of complex geometries and adaptation of multifunctional structures. Also, the tangible achievements of printed electronics, another additive approach to the manufacturing of electronics, prove the need for the development of new formats of manufacturing and present possible directions for going one step further than planar printed circuits. New perspectives from the structural electronics concept to integrate various materials and elements into multifunctional constructs enable the creation of unique electronic devices. That mixture of functional materials, freeform digital designs, new equipment and tailored postprocessing techniques can be enablers for the fabrication of active structures with electrical, mechanical, photonic or biological properties (Figure 6.1). In fact, it was just a matter of time that with such high potential technology in our hands, additive manufacturing techniques will be implemented for fabricating elements and circuits in electronics technology, building on the achievements of printed electronics. In this section, we will look closer at the current achievements in the synergistic integration of electronics with additive manufacturing to fabricate 3D-printed structural electronics.

The obvious drawback of well-established printed electronics technology is the application of patterning methods strictly dedicated to the formation of 2D conductive traces. Besides advances in manufacturing processes, the final product remains almost identical in function compared with conventional electronics based on PCBs. Further miniaturization and higher level of integration of electronic devices will rely on the adaptation of free-form fabrication of conductive and functional structures. This will be possible thanks to the customization of device geometries, the integration of hierarchical geometries, personalized constructs and prototypes. And this is where 3D printed electronics emerge. The general concept of additively manufactured structural electronics with structural bulk circuitry embedded inside casings or construction elements is quite simple to understand. The 3D multilayered circuitry is fabricated by selectively applying conductive and insulating (also construction) material for each layer, with additional conductive interlayer connections and optional mounting of nonprintable discrete electronic components or small subsystems line ICs [305, 310–312, 416]. And the same as for the printed electronics, 3D printed structural electronics is not expected to fully surpass traditional manufacturing methods, but to break the limits of current manufacturing techniques, promote higher integration and foster new formats of electronics and new functionality of surrounding us items. Fully 3D printed electronics is slower to manufacture than described earlier in-mold electronics systems, but with un-

https://doi.org/10.1515/9783110793604-006

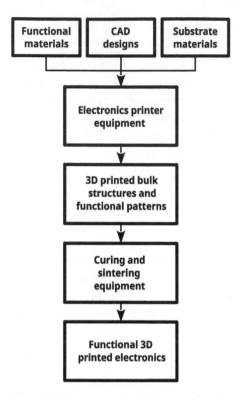

Figure 6.1: Key processes in additive manufacturing of structural electronics.

matched customization possibilities, it has bigger potential to fabricate tailored shapes and features for optimal functionality.

Many successful attempts to fabricate 3D printed electronic components and devices demonstrating the potential of additive manufacturing are presented in the literature. Most of them describe attempts to fabricate conductive paths, electrodes, circuitry, passive components or antennas, with a significant amount of experiments devoted to sensors, but at the same time less attention is toward more advanced active components or photonic and energy devices. Here, readers need to be advised that the term "3D printing" is consequently overused in the literature for the description of planar printing techniques, even with respect to printing on flexible substrates like paper, polymer films and textiles using screen-printing or other flat-bed mask techniques. Most often, "additive manufacturing" is used as a vast umbrella term covering almost all printing techniques used for fabricating electronic components. Indeed printed electronics is, in fact, also an additive approach, which was explained in previous sections. Therefore, this might be the source of misunderstanding. Moreover, it needs to be noted that for a single layer the fabrication of conductive paths in structural electronics is quite similar to printed electronics. In this book, however, printing techniques fabricating as much as possible spatial 3D objects (according to the mentioned ASTM definition) are described, with vertical interconnects, bulk load-bearings or spatial components, or at least conformal prints on 3D substrates and examples of achievements of 2D printed electronics techniques, exhibiting great potential for implementation in 3D printing of electronics (like inkjet adaptable for PolyJet). As a glimpse of the most revolutionary concepts introduced in the literature, there needs to be mentioned a fully 3D printed loudspeaker [417, 418], one of the most complete 3D electronics printers introduced to the market utilizing low-temperature, cured silver inks exhibiting resistivity close to bulk silver [419, 420]. 3D printed composites for fabrication of piezoresistive and capacitive sensors [66, 329], large-scale composite magnets [176] and highly-loaded with dielectric composites [421] with a portfolio of 3D printed resistors, inductors and capacitors and wireless "smart cap" for a milk package [422]. One of the first attempts to fabricate functional circuits such as a system based on 555 timer, electronic gaming die, a set of sensors embedded in a helmet [304, 423, 424] or low orbit satellite electronics [425] were utilizing a multiprocess approach based on SLA printers, DIW deposition of conductive traces, CNC milling machine and 6-axis robotic arm for efficient fabrication [425, 426].

Regarding the technical aspects and advantages of additive manufacturing, for most electronic components (single elements), there are two key processes: the printing process (deposition) and the optional curing/sintering process. The second one is obligatory for different types of ink-based materials (i. e., for evaporation of solvent) while for solid composites it is mostly not required. Compared to conventional planar electronic devices on rigid substrates, manufactured with photolithography, plating or vacuum deposition, this is a significantly simpler fabrication process (Figure 6.2). In the conventional approach, subtractive fabrication methods are generally planar, impeding the fabrication of complex geometries, voids, overhangs or 3D structures. What is more, such methods process a single material per processing step, which limits the production of electronics with multifunctional properties. Thanks to the maskless, additive approach, 3D printed electronics can also enjoy cost savings due to lower capital expenditure [214]. Other most visible advantages of 3D printing of electronics include the introduction of in-house prototyping of PCBs eliminating time-bottlenecks in prototyping, which takes weeks for PCBs and for all needed modification, and additionally safeguarding intellectual property, printing on delivered conformal surfaces favoring full optimizations of available space for slimmer product designs such as antennas and sensors onto precast plastic covers, lower material wastage typically high for subtractive processes, efficient streamlining of product development by fabricating electronics together in a single manufacturing process with construction elements and mass customization with ease of changing designs without the additional tooling costs or preparations [121, 353, 427].

Thanks to the introduced advantages, additive manufacturing of electronics is a rapidly expanding sector in the industry. Additive manufacturing of the electronics sector is generating around $1 billion US dollars of revenue, 13 % of the total additive manufacturing industry revenue and the 3D printed electronics market is expected to grow exceeding $3.5 billion US dollars by 2032 [428]. The biggest drivers here are electronics and aerospace industries interested in the fabrication of conformal functional antennas

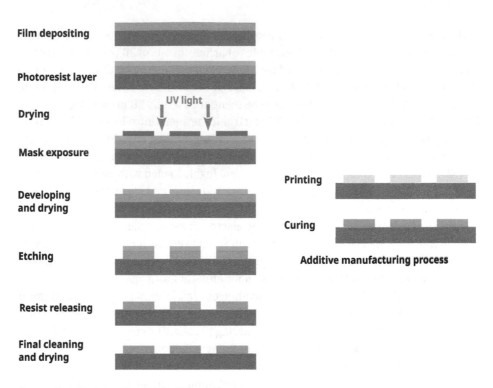

Film depositing

Photoresist layer

Drying

UV light

Mask exposure

Developing
and drying

Etching

Resist releasing

Final cleaning
and drying

Conventional substractive fabrication process

Printing

Curing

Additive manufacturing process

Figure 6.2: The comparison between conventional subtractive fabrication of printed circuit boards and innovative approach to additive manufacturing [228].

and sensors to reduce weight and space utilization for better fuel efficiency and aircraft performances or reduce footprints of electronic devices for higher integration [121, 429].

6.1 Conductors and passive components

Conductive paths are the most important part of any electronic circuit. In the area of conductive materials for additive manufacturing, the most visible is the development of filament composites for FDM (with metal powders, carbon black, nanotubes and graphene), metal pastes (mostly silver) for Direct Write and metal inks (also mostly silver) for Aerosol Jet Printing and inkjet techniques. Individual reports deal with powder-based techniques (metal powders) and UV-cured materials, but a large group of publications describes a hybrid approach for the fabrication of spatial 3D objects with conductive elements deposited with various techniques (sometimes not printed).

The greatest number of publications on 3D printed electronics research is by far in the area of electrically conductive and resistive composites for the most efficient and popular technique FDM (Figure 6.3). Due to the low loading needed to obtain per-

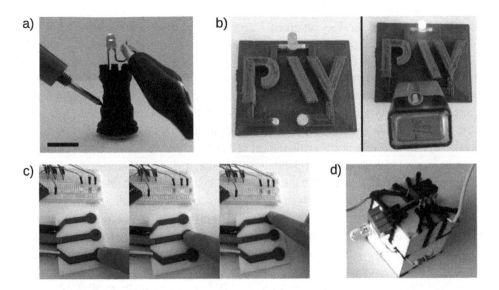

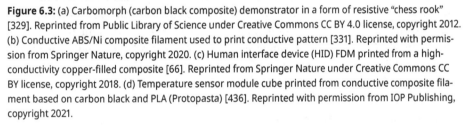

Figure 6.3: (a) Carbomorph (carbon black composite) demonstrator in a form of resistive "chess rook" [329]. Reprinted from Public Library of Science under Creative Commons CC BY 4.0 license, copyright 2012. (b) Conductive ABS/Ni composite filament used to print conductive pattern [331]. Reprinted with permission from Springer Nature, copyright 2020. (c) Human interface device (HID) FDM printed from a high-conductivity copper-filled composite [66]. Reprinted from Springer Nature under Creative Commons CC BY license, copyright 2018. (d) Temperature sensor module cube printed from conductive composite filament based on carbon black and PLA (Protopasta) [436]. Reprinted with permission from IOP Publishing, copyright 2021.

colation threshold, and thus the possibility to fabricate highly loaded composites the most promising and most investigated are nanocomposites, especially with carbon nanomaterials. Polymer thermoplastic composites in filament form containing carbon nanotubes [430, 431], graphene [432, 433] and carbon black [329, 434] usually have fairly high resistivity. Carbon-based composite filaments and FDM printed elements exhibit electrical conductivity values of 10^{-3} S/m to 10 S/m for loadings ranging up to 10 wt.% [329, 432, 435, 436], which correspond to several orders of magnitude higher values that for metals or even metal-filled composites. Such values are acceptable for the fabrication of resistors and other functional structures like sensors or electrostatic and electromagnetic shielding but are unsuitable for high-efficiency electrical circuits. Therefore, other approaches are explored, incorporating a metallic filler. One of the first literature reports, describing highly conductive FDM composites consisting of copper microflakes and the ABS polymer matrix presents a few orders of magnitude higher conductivity (10^4 S/m) than for carbon-based composites [66]. Also, addition of nickel micropowders to an ABS polymer was examined, and materials exhibited conductivity in the range of 10^2 S/m [331]. Such values of conductivity are typical for polymer composites with metal additives, and values of 10^3–10 S/m are often reported for such composites [437, 438]. Besides relatively high resistivity, despite the possibility of fabricating paths with large cross-section areas, another limitation of this technology is also the low current carrying capacity. The issue with high-loaded FDM composites is that they are often brittle and difficult to process with the typical off-the-shelf printers. Nevertheless, such materials are very popular due to the availability of commercial composite filaments and rather simple methods of fabrication with tailored functional fillers. Therefore, in the literature, many examples of functional demonstrators are presented. An example of implementation of FDM technology concerned a series of sensors fabricated from a carbon-polymer composite, including piezoresistive sensors utilized in a flex sensor embedded into a 3D printed glove sensing the flexing of a hand, embedded capacitive sensors detecting liquid amount inside in smart vessels and a capacitive sensing Human-Interface Devices (HIDs) tactile pad [329]. Later this material was commercialized as "Carbomorph" material—a polycaprolactone composite with 15 wt.% of amorphous carbon black phase. A tactile pad was also demonstrated with the Cu-ABS composites connected to an Arduino microcontroller [66].

While conductive composites tailored for FDM printing usually exhibit relatively high resistivity values, not suitable for most high-efficiency electronics, other options are also evaluated for the fabrication of paths and elements with conductivity matching bulk metals. While for FDM composites, the dominant role plays polymer dielectric based exhibiting high viscosity, the idea is to substitute the polymer matrix with less viscous and less dense vehicles and incorporate in it a larger amount of conductive particles. With substantially little modification to the deposition technique, such an approach can be achieved with direct ink write and corresponding pastes and inks. With the application of conductive pastes based on micro and nanoparticles of metallic or carbon nature, conductive/resistive elements can be fabricated (Figure 6.4). Such an approach is known for planar printing techniques and for the adhesives and sealants deposition, and instead of thermal melting of a composite, the paste is extruded from the nozzle in a room temperature process, additionally cured by the evaporation of the solvent. As an analogy to the FDM composites, also here conductive compositions with GNPs or CNTs with loading as high as 40–60 wt.% are often demonstrated, resulting in the electrical conductivity values reaching up to 10^3 S/m, which accounts for almost five orders of magnitude better conductivity comparing to the FDM polymer composite filaments [99, 439–442]. But the DIW technique has more advantages, especially in the use of metal powders as inks and paste conductive fillers. One of the first attempts to fabricate 3D printed electronic circuits was based on the conductive silver pastes deposited onto SLA 3D structures, reaching conductivity values up to 10^5 S/m [304, 424], which is comparable with analogous systems fabricated with screen-printing techniques on flat substrates [443]. This solution can be easily adapted for the most basic fabrication of conductive paths with relatively low resistivity. To achieve even better electrical properties, we need to prepare as many solid metallic materials as possible, contrary to composite pastes, and here metal nanopowders are the key solution. Similar to the use of carbon nanomaterials, the high loading of nanoparticles can be obtained for metal

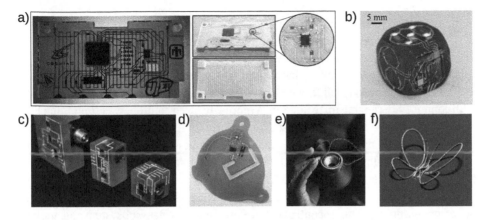

Figure 6.4: (a) 3D-printed CubeSat module produced by using SLA (left) and FDM (right) techniques, CNC routing and DIW [425]. Reprinted with permission from Springer Nature, copyright 2014. (b) SLA and DIW printed gaming die with embedded microcontroller and accelerometer, and LEDs on the surface [460]. Reprinted with permission from Springer Nature, copyright 2012. (c) Three generations of a three-axis magnetic flux sensor system fabricated with SLA and DIW [423]. Reprinted with permission from Solid Freeform Fabrication Symposium. (d) Accelerometer system helmet insert [423]. Reprinted with permission from Solid Freeform Fabrication Symposium. (e) Functional 3D printed loudspeaker with combined FDM and DIW techniques (author Jason Koski, reprinted with permission from Cornell University Photography) [417]. (f) Laser-assisted DIW fabricated 3D metal architectures [459]. Reprinted with permission from National Academy of Sciences.

nanopowders, yet they require additional thermal sintering to obtain almost solid metallic lines. Thankfully, this sintering process can be conducted in relatively low temperatures (even below 200 °C), which allows fabrication on polymer substrates. With the use of silver NPs and loading as high as 60–80 wt.%, electrical conductivity of 10^7 S/m is achievable [78, 444–447] comparable to the values of bulk silver or copper. With the DIW technique, other materials can also be utilized, such as MXenes, conductive polymers or compositions of multiple functional phases. With MXenes various microelectronic, radio-frequency, sensing and biological inks were prepared for the DIW printing of electronic structures reaching conductivity up to 200 S/m [448–452]. PEDOT:PSS hydrogel ink allowed the fabrication of conductive paths reaching 10^4 S/m [453], and a mixture of GNPs functionalized with Fe_3O_4 nanoparticles, additionally thermally annealed, exhibited conductivity of 580 S/m and in additionally a saturation magnetization value of 15.8 emu/g [454]. An unusual attempt is also the opposite of a heated solvent evaporation, a freeze drying, applied for 3D printing of conductive silver NW, graphene aerogels and even liquid metals [98, 453, 455–457]. Other techniques employ printing of graphene based pastes directly into a support gel [330]. While the majority of the applications of DIW are dedicated to fabricating conductive lines on the nonplanar and 3D printed substrates delivered by other techniques, like conformal spiral antennas based on nanocellulose and silver nanowires ink [458], this approach also allows to fabricate spatial 3D structures without additional scaffolds. The most simple examples are 3D coils or microgrids [459], but spacial conductive structure from graphene-polymer composite in the form of scaffolds for electronics application and proliferation of biology cells for tissue engineering was also fabricated with the commercial 3D BioPlotter (EnvisionTEC GmbH, Germany) [99]. A solution-based graphene paste composed of 75 wt.% graphene, 25 wt.% polylactide-co-glycolide was 3D printed under ambient conditions into scaffolds, with conductivity ranging up to 800 S/m. A more advanced solution was combining FDM and DIW fabrication processes were utilized to form a fully functional loudspeaker [417]. One printer (FDM) made the plastic cone and base of the loudspeaker, and the second printer (DIW) laid down the wires on the cone with silver paste, also creating a magnet ferrite ink. With the use of SLA and DIW techniques, a series of demonstrators were introduced including simple 2D circuits with 555 timer, electronic gaming die and magnetic flux sensor with a microcontroller, accelerometer, wireless subsystem and antenna embedded in helmet and a low orbit satellite control board [304, 423–425, 460]. A more advanced approach resulted in the spin-off Voxel8 with a complete system that uses DIW with developed conductive, particle-free, low-viscosity reactive silver ink based on $AgC_2H_3O_2$, curable at 90 °C, exhibiting an electrical conductivity of $>10^5$ S/cm, used for printing a simple circuit ended in the drone [420]. Quite a similar approach, but mostly for multilayer 2D circuits is the Voltera Circuit Board Printer with a propriety conductive and insulating inks [461]. In the microscale a high-resolution 3D printing of eutectic GaIn using a narrow-diameter glass capillary was used to fabricate features with a minimum line width of 1.9 μm [76]. Also, high aspect ratio microstructures were prepared of PEDOT:PSS, allowing to print pillar structures with 7 μm in diameter and 5000 μm in height [77]. A truly 3D printed conductive structure with 20 μm features DIW printed from Ag nanoparticle ink supported by laser sintering (a form of SLS) was also demonstrated with examples of metallic coils and interconnects, but also other forms of free-standing spiral architectures such as springs and butterfly shapes [459].

The next group of additive techniques covers ink-based printing, such as AJP or inkjet being adaptable to Multijet. While ink-based methods are often classified as 3D printing regardless of the type of produced structures, it is necessary to be precise, that even for spatial 3D objects, especially electronic elements and circuits, they actually still fabricate planar thin layers. Nevertheless, the possibility of adaptation and the level of introduction of ink-based techniques to 3D printing of electronics, even directly translated from printed electronics achievements is enormous for the most popular digital ink techniques such as inkjet and AJP (Figure 6.5). This is why here the examples of electronic elements fabricated with these techniques, even in the 2D formats, are intentionally presented, keeping in mind that the translation is achievable with less effort than for instance transforming screen printing to DIW even with similarities in used materials. The most significant advantage of these techniques is the ability to use metal nanopowders with a sintering process for low-resistivity elements, which is impossible for the FDM process. Also, a higher loading of carbon nanomaterials can be obtained due to the lack of excessive polymer matrix, obligatory for the FDM process or even for DIW. With these two ink-based techniques, conductive lines printed from metal nanoparticles and

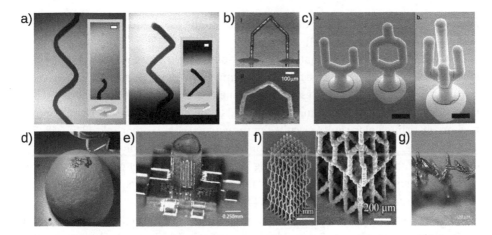

Figure 6.5: Various 3D microstructures of metal NPs fabricated by modified inkjet printing such as (a) micro zigzag, (b) metal micro bridge interconnector [475] (Reprinted with permission from IOP Publishing, copyright 2010) and (c) high-aspect ratio structures [477]. Reprinted with permission from John Wiley and Sons, copyright 2014. (d) AJP printed conductive logo on an orange [471]. Reprinted with permission from American Chemical Society, copyright 2021. (e) AJP printed vertical metal lines onto the wall of the dielectric pillar [472]. Reprinted with permission from IOP Publishing, copyright 2015. (f) AJP 3D printed microarchitectures [481] (Reprinted with permission from Elsevier, copyright 2021) and (g) microstructures [478]. Reprinted from AccScience Publishing under Creative Commons Attribution 4.0 International License, copyright 2023.

sintered in elevated temperatures allow to achieve, with little effort, conductivity values close to the bulk metals [131, 133, 462, 463]. Carbon nanomaterial- based inks allow fabrication of electrodes with conductivity only a few orders of magnitude lower (up to 10^4 S/m), which is the predicted limit for macrostructures made from carbon nanomaterials [127, 211, 354, 464–466]. And that would be all regarding the adaptation of ink-based techniques for printed conductive lines, while this topic is widely described in the literature for printed electronics applications. Not far from printed electronics, ink techniques are capable of fabricating multilayer, non-planar circuits with little modifications in the process [140, 467, 468]. Such an attempt is commercialized by the Nano Dimension company with its Multijet Dragonfly system for structural electronics [469]. Also, the AJP technique has been optimized from the beginning for the fabrication on nonplanar substrates, and the commercial equipment, such as provided by Optomec, is often capable of printing with six degrees-of-freedom [470]. This offers the possibility to print various elements inside the casings used commercially by several companies to print antennas for LTE, NFC, GPS, Wifi, WLAN and BT, with performance comparable to other production methods or more spectacular examples of printing on an eggshell, golf ball or fruits [471, 472]. The most interesting part is the preparation of various conductive inkjet printed 3D microstructures (bridges, pillar arrays, helices and zigzags) formulated from metal nanoparticle-based inks (silver or gold) with the conductivity reaching 10^6 S/m [473–478] and the same applies to the fabrication of 3D microlattices with the AJP technique [479–481].

Speaking of low-viscosity materials for ink-based techniques, a highly explored region in that scope is particle-free metallic-organic decomposition inks (MODs). As mentioned earlier, these materials resolve the problem of nanoparticle agglomeration and possible nozzle clogging. These inks can be used to metallize contacts, electrodes for capacitors or printed electronics circuits the same way as with other inks [340, 348, 351, 482]. Moreover, these inks can be successfully used on low-cost polymer substrates, additively manufactured [483]. The resulting layers are also of very good quality, as evidenced by their use in the fabrication of antennas [346], high frequency circuits coils [484] or electrodes of thin film transistors [485].

A more exotic additive manufacturing technique incorporated for the fabrication of 3D printed electronics is SLS, a powder bed approach. For electronic elements printed with the SLS (or SLM) technique, the approach is almost identical as for the fabrication of load-bearing elements, because a large group of materials used for these techniques are metal powders forming electrically conductive metal elements. A good example is the fabrication of copper and aluminium alloy horn antennas with SLM with parameters comparable to analogous solid metal antenna (Figure 6.6(a)) [486]. The same approach with aluminium alloys was introduced for the fabrication wideband multilayer waveguide array with an integrated corporate-fed network in the Ku-band with the DMLS technique [487]. Also, weight reduction was possible for the waveguide structures while maintaining their electrical properties with laser sintering of metal powders [488]. For more precise micro-components fabrication, silver and copper NPs were used instead of micropowders, resulting in electrical conductivity of 10^6 S/m, but due to the intensive agglomeration of nanopowder it is impossible to deposit them in the powder-bed manner and other techniques, such as DIW, needs to be incorporated [459]. With the SLS technique, conductive composite powders can also be processed. Conductive 3D objects CNTs/TPU based (flexible thermoplastic polyurethane), exhibited good flexibility and electrical conductivity of 0.1 S/m [489], and a similar composite was introduced for the fabrication of a 3D flexible piezoresistive sensor [490]. A composite powder with iron oxide nanoparticle-coated polyamide powder (PA12) was used for the fabrication of print magnetic parts [491] and magnetic field activated grippers were also prepared by SLS from TPU powders filled with micron-sized Nd-Fe-B magnetic particles [492].

Having such enormous possibilities to fabricate conductive electrodes we are able to focus on the development of another passive element, a capacitor. In many reports, fully printed capacitors are demonstrated prepared from silver and gold nanoparticle inks with the planar inkjet technique [212, 493], and naturally with carbon nanomaterials conductive filament on FDM [494], exhibiting capacitance values 3.1 nF cm^{-2}, 314 pF and 150 pF, respectively. And while inkjet is a very close technique to Multijet, also multilayered capacitors and inductors were fabricated with inkjet printing of AgNP inks [495]. While the special interest regarding the application of additive manufacturing

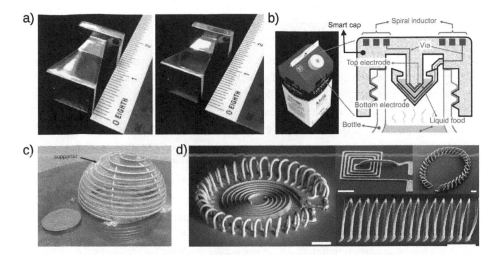

Figure 6.6: (a) Metallic SLM 3D-printed k-band-stepped double-ridged square horn antennas [486]. Reprinted from MDPI, Basel, Switzerland, under Creative Commons Attribution 4.0 license, copyright 2027. (b) 3D printed "smart cap" for rapid detection of liquid food quality featuring wireless readout [422]. Reprinted from Springer Nature under Creative Commons CC BY 4.0 license, copyright 2015. (c) Electrically small spherical helix antenna using 3D printed polyamide material painted with a silver paste [500]. Reprinted with permission from John Wiley and Sons, copyright 2016. (d) Gigahertz electromagnetic structures via DIW for a radio-frequency oscillator and transmitter applications [501]. Reprinted with permission from John Wiley and Sons, copyright 2017.

and nanomaterials is focused on supercapacitors, they will be described in detail in the energy storage devices' sections.

Conductive structures are also a starting point for the fabrication of other passive components—inductors and antennas, being the crucial components of various systems in the fields of smart packaging, healthcare, anticounterfeiting protection and public safety (Figure 6.6(b)). Compared to conventional antenna manufacturing methods, 3D printing introduces more materials selection and allows for complex structural design. 3D printed free-standing conductive structures from CNT-PLA composites were formed as a 3D helix (inductor) [496]. Other attempts were focused on FDM printed inductors for wireless charging applications [494] and printed antennas for radio frequency identification (RFID) tags [497]. Conductive composites with nanomaterials are also sufficient for the fabrication of lightweight electromagnetic interference (EMI) shielding from PLA/graphene with 16 dB reflection in the X-band frequency [498]. FDM was also used to print horn antenna from PLA, later metallized with commercial conductive spray [499] the same as a spherical helix antenna fabricated with SLS from PA12 and painted with silver paste (Figure 6.6(c)) [500]. Tu built a 3D printed loudspeaker described earlier in this book and a conductive coil that needed to be fabricated with the DIW technique and silver conductive paste [417]. With DIW also, micrometer scale lines (1.9 μm) allowed to print a 2D square coil with 3D interconnects [76]. Passive RF freeform structures printed with DIW and silver nanoparticles were fabricated as an inductor-capacitor (LC) resonator consisting of a 32-turn toroidal inductor and parallel-plate capacitor, with a line width of 10 μm, exhibiting a resonance frequency of 6.5 GHz (Figure 6.6(d) [501]). A 3D spiral antenna with DIW and silver nanowires ink was printed on the hemisphere surface, with a resonance at 2.48 GHz [458]. Electrically small antennas were also fully printed with inkjet both for plastic UV-curable support and silver conductor [467]. Obtained patch antenna exhibited a gain value of 8 dBi and radiation efficiency of 81 % at 2.4 GHz, matching the performance of the solid substrate antenna. Another electrically small antenna was 3D printed with silver nanoparticle inks, in the form of meander lines of about 100 μm on convex and concave hemispherical surfaces, working at 1.73 GHz with 71 % [502]. In the micrometer scale, an array of plasmonic structures forming vertical split-ring resonators (SRRs) were prepared from aerosol-printed gold nanoparticles [503]. Such fine detail structures (line width below 2 μm) exhibited a resonance at 19.5 μm wavelength demonstrating the possibility to be applied as an antenna. Finally, a fully 3D printed horn antenna, with mechanical properties matching the solid elements, was SLM printed from copper and aluminium alloys. Both have an impedance bandwidth across the K-band and exhibit gain parameters of 13.23 dBi at 25 GHz and 13.5 dBi at 24 GHz, respectively, proving that they can replace the conventional antenna [486]. Many other attempts at 3D printed antennas are presented in the literature, also as fully dielectric antennas, polymer lenses, polarizers or waveguides [504].

6.2 Active components

It is a natural course that the great majority of the reports presented in the state-of-the-art are focused on the conductive electrical components, being the fundamental building block of all electronics systems therefore the overwhelming size of the previous section. In the following sections, other electronic applications of materials and additive techniques will be briefly highlighted.

Active components such as diodes and transistors are the key components of modern advanced electronic circuits. Especially in the case of transistors, such active components are typically multilayered, built from different materials such as semiconductors, conductors and dielectrics. The most simple transistor consists of a semiconducting channel, electrodes (drain, source, gate) and isolation dielectrics (Figure 6.7(a)). On the mass scale, the structure is more complicated, especially for high-performance active components fabricated with silicon wafer technology. Yet, the research in printed electronics for at least two decades is focused on the large-scale fabrication of printed diodes [279], transistors [493, 505, 506] and even entire microchips [274, 507]. For the 3D printed electronics, the transition from the printed electronics achievements is not so simple and direct. The trick is in the structure of such transistor, while the main operation principle is based on the thin layer structures, and currently none of the truly 3D printed techniques is capable of fabricating such fine features (in the range of μm), while this is not a goal for large scale processes. Of course, inkjet, DIW and AJP techniques that are also utilized in the fabrication of 3D structures are able to print such fine structures. Due to the planar nature of the transistor inkjet and AJP plays a significant role in their manufacturing, and the fabrication of high-efficiency active circuits needs the integration of 3D printing techniques such as FDM, SLA or SLS with ink-based techniques well developed for printed electronics. What is more, the technology and market analyses indicated from the early days of the adaptation of 3D printing for electronics, which is the most important element of modern electronics—3D printed transistor—might not be fully realized within decades [508]. Today these predictions are still valid, and thus in the literature we can find reports with titles containing phrases "3D printed transistor" or similar statements. They mostly utilize inkjet, DIW or AJP techniques for depositing channel and electrode materials [367, 509] even indicating that the TFT is fabricated on flexible substrates as in printed electronics, an occasionally and clearly presenting approach as a hybrid process involving inkjet printing [510]. On the other hand, individual reports deal with the fabrication of PEDOT-based composites with SLA (not the entire transistor structure) [511] and there is no technical limitation to print semiconductor materials with FDM, even in the microscale regime [512], but such approaches have not gained a lot of attention for fabricating 3D printed transistors. This should not come as a surprise, while there is no objective reason for building macroscale transistor structures with additive manufacturing, especially when the research result indicates that the performance of transistors, such as the ON/OFF ratio parameter, is compromised by a larger geometry of the channel [367]. Interestingly, it is worth mentioning that the high-performance silicon chips in the nm technology are also developed in the direction of 3D structures, with the channel surrounded by the dielectric material of the gate like Fin-FET or Gate-All-Around FET (GAA FET) (Figure 6.7(b)) [513]. But also, in this case, readers need to be aware that these structures are still fabricated with photolithography-based silicon technology, and not with an additive manufacturing approach [514, 515].

As mentioned, there are no direct examples of 3D printed transistors, due to the nature of such structures, but some of the latest results incorporating techniques adaptable for 3D printing are worth mentioning here. The most explored materials used for the fabrication of printed transistors are semiconducting polymers, inorganic semiconducting oxides and nanomaterials (Figure 6.7(c, d)). The most basic approach covers the use of the most popular materials, PEDOT:PSS, achieving the carrier mobility μ values of 0.437 cm^2 V^{-1} s^{-1} [516], which sounds not very impressive compared to the high-performance silicon transistors. Other materials allow improving this value by the order of magnitude with other organic materials such as dinaphtho[2,3-b: 2',3'-f]thieno[3,2-b]thiophene (DNTT), or 2,8-difluoro-5,11-bis(triethylsilylethynyl) anthradithiophene (diF-TES-ADT) achieving 1.1 cm^2 V^{-1} s^{-1} and 6.7 cm^2 V^{-1} s^{-1}, respectively [517, 518]. Switching material to inorganic semiconductor oxides allows to significantly improve the performance of printed transistors, reaching 19 cm^2 V^{-1} s^{-1} [519], and even 230 cm^2 V^{-1} s^{-1} for ITO printed inks [520]. Semiconductor oxide-based inks allow for the fabrication of 3D printed arch-type, cylindrical architectures with the DIW technique,

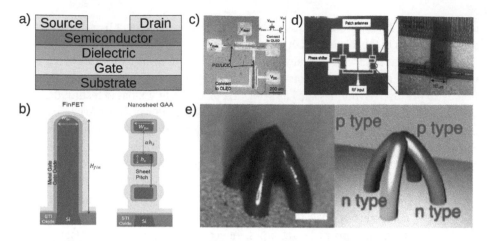

Figure 6.7: (a) Schematic illustration of the cross-section of thin-film transistor. (b) Comparison of FinFET (left) and GAA FET (right) cross-sections [513]. Reprinted from MDPI, Basel, Switzerland, under Creative Commons Attribution 4.0 license, copyright 2022. (c) Control circuit made of two fully inkjet printed top-gated SWCNT TFTs [505]. Reprinted with permission from American Chemical Society, copyright 2011. (d) Photographs of the AJP printed SWCNT-based flexible device with printed TFTs [526]. Reprinted with permission from Elsevier, copyright 2010. (e) Photograph and illustrated models of arch-type architectures consisting of junctional p- and n-type inorganic semiconductors [522]. Reprinted with permission from Springer Nature, copyright 2021.

with p-type and n-type electrodes from $Bi_{0.5}Sb_{1.5}Te_{3.0}$ and $Bi_{2.0}Te_{2.7}Se_{0.3}$, respectively, thus not used as a transistor but as a thermoelectric generator [521]. A similar approach was also used for fabrication of 3D thermoelectric structures with bipolar junctions (Figure 6.7(e)) [522]. Nanomaterials are also extensively evaluated as a material for the channel of the transistor. Carbon nanomaterials play an important role here and they allow to easily achieve comparable performance to the organic semiconductors with a less complicated procedure of preparation. Examples of printed graphene and carbon nanotube transistors are reaching mobility values in the range of 0.23 to 30 cm^2 V^{-1} s^{-1} [505, 523–526]. Also, incorporation of MXene inks allowed to fabricate TFT exhibiting carrier mobility value of 0.5 to 2.61 cm^2 V^{-1} s^{-1} [527]. Regardless of the materials used for the fabrication of transistors, printing techniques allow fabricating more complex active structures such as logic gates or memory blocks [363, 528, 529]. In the end, it is worth to mention the PlasticARM, a fully functional 32-bit arm microprocessor, fabricated in cooperation of Arm Ltd and PragmatIC Semiconductor Ltd, Cambridge, UK, built as Cortex-M processor, RAM and ROM [530]. Although this is not a printed structure, fabricated from n-type IGZO on a 200-mm diameter polyimide wafer using conventional semiconductor processing equipment, it is still a very impressive and inspiring achievement, setting the directions for the fabrication of plastic electronics.

6.3 Photonic devices

Photonic devices are another class of electronic elements that benefit from the adaptation of additive manufacturing. In their nature, photonic devices either produce, manipulate or detect light, and the most known examples of such structures are light-emitting diodes (LEDs), various forms of other light sources and displays and photovoltaic cells (PVs). Beside such active components as LEDs, also passive photonic microstructures are implemented, consisting of arrays of micro and nanostructures modulating the light beam. In the area of photonic devices, there is more space for adaptation of additive manufacturing than for printing transistors, while not all materials and structures need to exhibit exceptionally high electrical conductivity or carrier mobility. But similar to active components, also for photonic devices fabricated with printed electronics, the majority of printed structures are constructed of thin and thick films printed on the substrate. This also limits the capability of creating truly spatial 3D objects, and also here there is no reason for producing such components, while emission or absorption of light is mainly a surface-related phenomenon.

The large group of materials explored for the printing of emissive layers are semiconducting polymers, and the devices fabricated with them are called organic light-emitting diodes (OLEDs) or organic photovoltaics (OPVs). Such components often consist of several layers on top of each other, including backside electrode, electron transport layer, active layer, hole transport layer and transparent electrodes, thus the combination might vary [531–533]. Besides, often used coating techniques and also vacuum deposition for the fabrication of OLEDs, the most often used technique is inkjet even for the fabrication of 8K 55″ display [533, 534]. On the other hand, inkjet techniques can be exploited for the fabrication of precise contact electrodes or transparent electrodes for light-emitting devices [535]. Also, many types of hybrid approaches are introduced for the fabrication of OLEDs, such as printing Ag NPs and PEDOT:PSS, with AJP and DIW techniques (Figure 6.8(b)) [325]. But besides organic materials, quantum dots (QDs) are also used as direct emissive layer or luminophore. Here, also inkjet printing is utilized with various materials such as CdZnSe/ZnS core-shell QDs [536], PbS-CdS core-shell QDs [537] or $CsPbBr_3$ perovskite QDs [538]. One of the most known approaches utilizing the DIW technique is a report describing the deposition of a 5-layer LED with an emissive layer made up of quantum-dot material printed on a contact lens, achieving light emission comparable to LEDs printed on flat substrates (100 cd/m^2) (Figure 6.8(a)) [323]. Application of DIW eliminates the occurrence of the coffee-ring effect, one of the main problems associated with inkjet printing, not very problematic for metallic conductive paths but significantly impacting the efficiency of printed LEDs, introducing high surface roughens in printed layers [539]. There are also attempts to print perovskite materials, including perovskite quantum dots, even for the application of printed lasers, which can be beneficial for structuring natural resonant cavities and amplified spontaneous emission [540, 541]. A multinozzle DIW printing allowed the fabrication of integrated light-emitting and light-detecting 3D structures, where light-emitting/detecting parts were made from multiple metal, dielectric and semiconductors inks printed through separate nozzles simultaneously, later applied for inspection of structural defects in a wing of a model airplane [542]. A DIW-printed quantum dot nanophotonic ink was also used to fabricate RGB pixels for displays in the form of 3D nanopillar structures with increased brightness for high-resolution display devices [543].

A sort of opposite structure to LEDs are photovoltaic cells and photodetectors. For the thin film OPVs, their structure is often similar to OLEDs. Also here, inkjet is used for the fabrication and perovskites are employed for printed PVs [544–547]. Some reports suggest that printed solar cell devices might outperform spin-coated ones [545]. In addition, the 3D printing approach offers control over the crystallization behavior of perovskite layers, a critical parameter for the performance and lifetime of PVs, compared to other coating techniques [548, 549]. A halide perovskite solution was even used to print a 3D structured X-ray photodetector with the AJP technique [550]. AJP was also successfully utilized for the fabrication of active layers [551] and current collecting grid electrodes in solar cells and photodetectors [552–554].

One of the biggest potentials in additive manufacturing for photonic applications is in the 3D fabrication and deposition of composite luminophores and scintillators (Figure 6.8(c, d)) [555, 556]. The key part plays the processability of quantum dots here in a colloidal solution, creating an opportunity to prepare materials highly compatible with Multijet and DIW 3D printing processes [557, 558]. Almost all inks for inkjet printing with QD nanomaterials that are mentioned in the literature can be directly used in the Multijet 3D printing technique with no or little modifications to their composition. What is more, while luminophorous composites do not have to be electrically conductive, adaptation for SLA to build optical layers and structures is promising. VAT techniques have already been used for the demonstration of several optoelectronic structures, like titanium dioxide NPs resin composites for tunable refractive index [559, 560], CdSe-ZnS fluorescent QD composites [561] and a low-cost PET scanner fabricated with 3D printed plastic scintillators [556].

The last area of photonic applications is the fabrication of metasurfaces and metastructures. A modified AJP technique was used to fabricate 3D freeform plasmonic structures such as vertical split-ring resonators made of gold NP inks [503]. A solvent-based ink for DIW was used to print complex photonic patterns on heated silicon substrate [562]. Also, DIW was used to print perovskite nanowires to fabricate photonic devices with digitally programmable polarization anisotropy, allowing the programming of optoelectric devices, such as RGB displays, optical strain sensors and optical information storing devices [563]. One of the most impressive results of micropatterning, including photonic metastructures are obtained with the two-photon lithography technique, enabling fabrication of both polymer and metallic micro and nanostructures, also microoptical structures, such as prisms, Fresnel lenses and microgratings [564]. And while

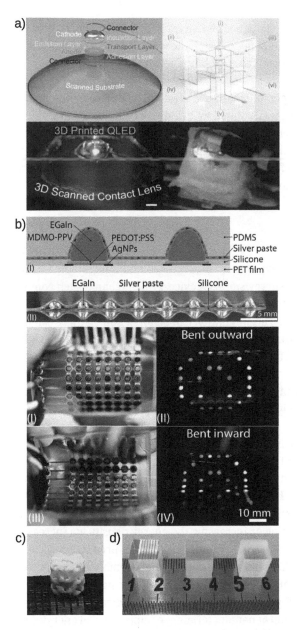

Figure 6.8: (a) 3D printed quantum dot light-emitting diodes onto a curvilinear contact lens and embedded in the cubic structure [323]. Reprinted with permission from American Chemical Society, copyright 2014. (b) 3D printed flexible OLED displays with the demonstration of different combinations of bending orientations and a cross-section of structure [325]. Reprinted from American Association for the Advancement of Science under a Creative Commons Attribution License 4.0 (CC BY), copyright 2022. (c) 3D-printed inorganic polycrystalline scintillator [555]. Reprinted with permission from Royal Society of Chemistry, copyright 2017. (d) DLP 3D-printed plastic scintillators for dosimetry applications [556]. Reprinted with permission from Elsevier, copyright 2022.

such a technique has its limitations and there is often a tradeoff between the largest object that can be printed and the minimum feature size, the work is on the way to achieve both fine microstructure architectures (1 μm) with large spatial dimensions (up to centimeters) [565].

In the end, while not directly photonic devices, also conductive components are incorporated for electrooptical devices fabricated with additive techniques. A combination of passive conductive components and photonic applications leads to the fabrication of precise microstructures such as DIW printing with graphene ink to formulate honeycomb pillar arrays as support for discrete LED for the fabrication of flexible light-emitting displays [566] or metal connectors for LED arrays made with AJP [479].

6.4 Energy storage devices

Another important direction in the development of advanced and integrated electronics is the area of high-performance energy storage devices. On one hand, such components retain energy based on the electrochemical reaction (batteries) and on the other are based on the charge separation and accumulation (capacitors). Both of these components share similarities in their construction, and they are mostly built of two electrodes (anode and cathode), additional electrolytes and a separator (for batteries and supercapacitors). In the last decades, nanomaterials such as graphene and carbon nanotubes have been extensively adapted for the fabrication of the electrodes in such energy storage devices [567–569], and we already know that nanomaterials have great potential for the application in additive manufacturing. In most of the applications, the shape of energy storage devices is standardized in a circular or rectangular structure, optimized for mass production. With the advent of IoT or wearables battery or capacitor is expected to become smaller or adopt alternative form, not to occupy most of the device volume, but maybe more efficiently utilize the space as adaptive, free-form structures. Again, the transformation from 2D to 3D structure might allow the fabrication of smaller or size adaptable energy devices retaining or increasing the energy density via increased volumetric usage of the electrodes [155, 570, 571]. Additionally, the production cycle of the battery can be optimized with 3D printing minimizing the assembly process, using less materials, and shortening the manufacturing time, thus lowering the total cost of the process [572]. Finally, combined geometry modification and electrodes integration via 3D printing and effective utilization of nanomaterials can improve areal energy density with a smaller footprint [573]. While the scope of the device for energy generation, storage and transformation fabricated with additive manufacturing is much broader than batteries and capacitors, including mentioned already photovoltaics, but also fuel cells and all sorts of solid-state generators and energy harvesters (thermoelectric, radio-frequency, tribo and piezoelectric) [574–577] here only the application of small batteries sufficient for powering IoT devices or capacitors for signal processing circuits will be covered.

An example of successful fabrication of a 3D printed battery is a free-standing lithium-ion battery with all components printed with the DIW technique, using water-based graphene oxide paste for the battery electrodes and a solid-state electrolyte containing Al_2O_3 nanoparticles, exhibiting a capacity of up to 170 mAh/g [578]. The DIW technique has many advantages in the fabrication process of batteries and supercapacitors, enabling the use of high-filled pastes with various conductive and supporting materials or electrolytes. Using water-based lithium titanate $Li_4Ti_5O_{12}$ (LTO) and lithium iron phosphate $LiFePO_4$ (LFP) nanoparticles, with up to 60 wt.% loading of nanomaterials 3D interdigitated microbattery architectures were printed with DIW, attaining high aspect ratios geometry of 3D electrodes (1:10) (Figure 6.9(a)) [579]. The addition of a standard liquid electrolyte ($LiClO_4$) allowed do achieve energy and power density reaching 9.7 J/cm^2 and 2.7 mW/cm^2, respectively. Lithium-based 3D printed batteries are a popular trend in the fabrication of 3D printed energy components. A combination of LFP and LTO fibers with PVDF for printing electrodes and CNT dissolved in gel polyelectrolyte NMP exhibited high specific discharge capacity and excellent flexibility [580]. Also, LFP paste for DIW with cellulose nanofiber (CNF) [581], and freeze-dried sulfur/carbon composite with Li-S [582] were used for 3D-printed battery fabrication. On the other hand, a miniature zinc-ion battery was demonstrated [583]. With the adaptation of 3D printing, the batteries with free-form structures were fabricated—a ring shape and capital letter H and U. Such water-based Zn-ion battery uses zinc ions (Zn^{2+}) as charge carriers instead of lithium ions reducing the risk of fire attributed to Li-on batteries, thus making them safer and more stable to atmospheric moisture. For the same reason, sodium-ion batteries were also 3D printed with the DIW technique [584].

DIW is highly utilized for the fabrication of electrodes with highly loaded nanomaterials, such as graphene hydrogel–polyaniline nanocomposites used in a supercapacitor [585]. Graphene-based aerogel pastes and inks were used in hybrid DIW and inkjet printing of electrodes with periodic macropores for the fabrication of supercapacitors with rate capability up to 10 A/g, powering a red LED and digital timer for several minutes, and a small electric fan for several seconds [586]. Many other reports of DIW printing adapted for the fabrication of 3D printed electrodes can be found in the literature, like thermo-responsive inks filled with graphene and copper used to fabricate the high aspect electrodes, additionally sintered, for electrochemical battery (Figure 6.9(b)) [587], CNT-based electrodes for microsupercapacitors [573] or molybdenum disulfide (MoS_2)-graphene hybrid aerogels for porous electrodes in sodium-ion battery [584]. Also, MXenes were printed with DIW and inkjet techniques for the fabrication of freestanding supercapacitors electrodes [588, 589].

Adaptation of 3D printing also allows the introduction of a controlled porosity in the electrode structure, enhancing the effective total area of the electrodes, with the remaining macroscale geometry. This is especially visible for structures with distributed pore sizes ranging from macro to nanoscale, what is called a hierarchical structure. Such devices typically exhibit enhanced performance compared to those with stochastic feature sizes, via enhanced charge transfer and diffusion of ions [590]. For example, high surface area nanoporous graphene-based material was used to print electrodes of a battery with pore sizes from 4–25 nm (through holes) to 500 μm (square pores) [591]. Supercapacitors based on hierarchical graphene aerogels with periodic macropores printed with DIW achieved superior rate capability compared to the other carbon-based electrodes [586]. Also mentioned above, sodium-ion battery anodes were fabricated with nanopores (300–800 nm in average size) [584] and DIW printed nanoporous graphene-based cathodes for the $Li–O_2$ batteries with improved active site utilization and ion transport capacity over the 2D filtration films [591]. Metallic porous electrodes were also fabricated with the AJP printing technique, as 3D structures in the form of porous microlattices of silver nanoparticles, used as electrodes for lithium-ion batteries [480]. The microlattice not only enhanced the electrical properties via a larger surface area influencing capacitance (by a factor of 4), but it also enhanced mechanical strain tolerance under frequent charge/discharge cycles.

Summarizing other techniques than highly utilized DIW, graphene-based PLA composite filaments for FDM were used to fabricate electrodes for the pseudo-capacitor [592] or PLA composites with submicron graphite for 3D printing of electrodes for electrochemical devices [593] and similar composites for solid-state supercapacitor fabrication [594]. Such examples are presented in Figure 6.9(c, d). Also, VAT techniques are utilized. A polymer composite containing silver NP decorated lead zirconate titanate powders (PZT) were mixed with UV curable resin to fabricate 3D printed capacitors. Due to the ability to impart a high polarization effect, the polymer nanocomposite exhibited a 30-fold increase in dielectric permittivity enhancing specific capacitance over similar thermoplastic composites (up to 63 F g^{-1}) [595]. Polyaniline-coated carbon fiber cathodes and solid zinc anodes along with an SLA-printed porous separator were used to fabricate rechargeable aqueous zinc-ion battery [583].

One of the unusual advantages that need to be mentioned for the application of additive manufacturing for energy devices is the ability to 3D print solid-state electrolyte materials. The solid-like electrolytes offer the advantage of limiting potential leaking or combustion like their fluid counterparts. DIW was used to print multisolid electrolyte paste containing $Li_7La_3Zr_2O_{12}$ (LLZ) garnet allowing the fabrication of a variety of freeform structures both conformal and a self-supporting structures [596]. All components of a lithium-ion battery using DIW were prepared, including water-based solid electrolyte paste composed of a highly concentrated graphene oxide sheet with optimal viscosity [578]. The PVDF-co-HFP-based paste-like with a Li ionic-liquid electrolyte was used to fabricate hybrid solid-state electrolytes using an elevated-temperature DIW technique [597] (Figure 6.9(e)). An all-component 3D printed lithium-ion battery was fabricated with DIW by printing graphene composite pastes containing LFP and LTO acting as electrodes and a solid-state gel-polymer electrolyte [578].

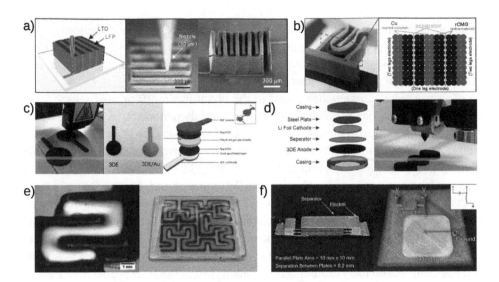

Figure 6.9: (a) 3D printed interdigitated Li-Ion microbattery architecture fabricated with the DIW printing technique [579]. Reprinted with permission from John Wiley and Sons, copyright 2013. (b) DIW printed supercapacitor device including a cellulose paper separator and positive and negative electrodes, with a scheme of the cross-section [587]. Reprinted with permission from American Chemical Society, copyright 2017. (c) Schematic illustration and 3D printed electrodes of solid-state supercapacitor fabrication, with adapted electrochemical plating technique [594]. Reprinted from Springer Nature under Creative Commons CC BY licence, copyright 2018. (d) 3D printed graphene-based energy storage device [592]. Reprinted from Springer Nature under Creative Commons CC BY license, copyright 2017. (e) 3D printed hybrid solid-state electrolyte Li-Ion full cell battery [597]. Reprinted with permission from John Wiley and Sons, copyright 2018. (f) Fully 3D printed high-pass filter LC circuit with 3D printed capacitor [494]. Reprinted with permission from Elsevier, copyright 2017.

6.4.1 Sensors

The measurement of environmental parameters as well as the life parameters of society members is gaining more traction these days, especially in the advent of IoT, industrial maintenance and personal electronics (wearables) integrating various types of sensors in one net for constant motoring of strain in the constructions, temperature of the body and in facilities, chemicals and biomarkers concentration in the environment and in our bodies [598–600]. The increasing need to apply sensors almost everywhere forces the adaptation of sensor structures for different environments or geometry of the objects. Here, structural electronics approach is a key player, and additive manufacturing can be easily used to form conformable sensors or embedded in 3D structures. The advantage of 3D printing increases the degree-of-freedom in sensor fabrication taking to the next level their integration for active monitoring. Due to the large variety of materials and structure adaptation for sensors (resistive, electrochemical, composite, conformal, etc.), one of the most broadly explored areas of research in 2D and 3D printed electronics is sensor applications. As in previous sections, the reports on sensor applications are presented with respect to the elements fabricated additively in 3D geometrical shapes or embedded inside 3D structures, leaving a large part of printed electronics achievements of sensors printed on flat and elastic substrates.

The most often combination found in the literature for 3D printed sensor fabrication is the adaptation of conductive materials, mostly nanomaterials, with DIW and FDM techniques. Utilizing graphene or nanotubes for preparation of pastes is one of the simplest approaches to apply these nanomaterials for 3D printed sensors (Figure 6.10(d)) [601, 602], and also with the use of an organic hydrogel from egg yolk [603]. Strain sensors printed from graphene-PDMS paste with DIW were prepared in the form of a mesh prepared through multiple printing [604]. The same material was later modified with silica nanoparticles to enhance the mechanical and thermal properties [605]. One of the most advanced demonstrators 3D printed in the form of multilayered scaffolds, spider and starfish-shaped objects were prepared with water-based MWCNT-chitosan compositions, additionally exhibiting self-healable properties after partial fracture (Figure 6.10(e)) [606]. Self-healable strain sensors with various shapes were also prepared with DIW in planar form with MXene and cellulose nanofibers mixture, with easy adaptation printing 3D structures [448]. A piezoresistive sensor with soft materials was prepared from carbon black-polyurethane composite, saturated with NaCl inclusion later removed by the water, creating an excessive porous structure with hierarchical porosity, demonstrating the ability to weigh small objects, measuring the dynamic vibration of a machine, monitoring pulse on the wrist, swallowing and blinking motions and also during a speech [607]. Besides, carbon nanomaterials and also metal-based pastes are used, such as submicrometer-sized silver particles dispersed in a highly stretchable silicone elastomer to form 3D printed stretchable tactile sensors directly printed on the human skin, detecting moisture, pulses and finger motions [608] and silver NW with cellulose nanocrystal compositions for eco-friendly disposable wireless sensor systems of chemical ion sensing [458]. Silver NW and DIW from previous experiments were also used with minor modifications to print deformation sensors conformally deposited on the surface of the porcine lung, for the spatial mapping of deformation measured by electrical impedance tomography [609]. In the area of low viscosity inks for sensors fabrication, there are also reports of AJP printed graphene sensors for immunosensing [610] or silver NP-based sensors for cerebral aneurysm monitoring [611], but as all low viscosity inks techniques they still have limitations for fabricating truly 3D printed elements.

While for most sensors a high electrical conductivity is not obligatory, even better for the measuring system if it is in the region of $10-10^6$ ohms, the great potential lies in the composite materials, especially for the FDM technique, being the most suitable for the fabrication of bulk structures with embedded sensors. Commercial composite filaments such as "carbomorph" and "blackmagic" are among the most often used for FDM fabrication of electronics, along with similar composites based on carbon micro and nanomaterials in PLA, ABS or any other thermoplasts [329, 612, 613]. Biomedical

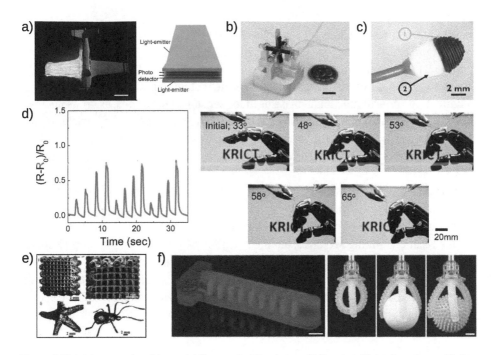

Figure 6.10: (a) Structured multimaterial filament for 3D printing of bifunctional light-detecting and light-emitting structure in detecting structural defects I aeroplane wing [542]. Reprinted from Springer Nature under Creative Commons CC BY license, copyright 2019. (b) 3D cubic cross-multiaxial force sensor on a FDM printed structure fabricated with CNT–TPU filaments [430]. Reprinted with permission from Elsevier, copyright 2017. (c) Three-dimensional-printed electrochemical sensor for simultaneous dual monitoring of serotonin overflow and circular muscle contraction [615]. Reprinted with permission from American Chemical Society, copyright 2019. (d) DIW printed nanotube and graphene composite strain sensors for detecting robot finger motions with different bending angles [601]. Reprinted with permission from Royal Society of Chemistry, copyright 2017. (e) Chitosan–CNT solvent composites for sensors, demonstrating 3D printed scaffolds as spider and starfish-shaped structures [606]. Reprinted with permission from Royal Society of Chemistry, copyright 2018. (f) Soft somatosensitive actuators via embedded 3D printing innervated with multiple soft sensors, and soft robotic grippers with somatosensory feedback [628]. Reprinted with permission from John Wiley and Sons, copyright 2018.

sensing systems can also be fabricated with FDM printed composites built from PLA matrix and carbon nanomaterials, and the biocompatibility of such materials is an important factor [614]. In other reports, a simple electrode printed as a cone from carbon-PLA composite was used as an ex vivo sensor implemented for the guinea pig gut tissue monitoring, allowing the measurements of serotonin level and muscle contraction (Figure 6.10(c)) [615]. Many other approaches describe graphene-PLA electrodes for H_2O_2 detection [616], mycotoxins [617], metal ions in biological specimens [618], glucose [619] and immunobiosensor tailored for the detection of viruses [620]. Graphene-ABS composites were also used to 3D print with FDM an electrocardiogram (ECG) sensor electrodes, additionally coated with metals (titanium and gold) to limit corrosion and oxidation [621].

Flexible thermoplastic matrix allowed to prepare MWCNT-polyurethane elastic strain sensors monitoring changes in materials properties via piezoresistive-effect measurements [622] and other polyurethane composites with MXene and MWCNTs were implemented for elastic strain sensors and electromagnetic shielding structures [623]. There is no limit on the type of functional phase used for sensing applications, as far as it allows the measurements of the properties change, therefore, magnetite nanoparticles (Fe_3O_4 NP) in ABS were used for the FDM printing of a glucose sensor [624] or gold nanoparticles (Au NPs) immobilized on the surface of printed PLA/graphene electrochemical sensors [625].

For the same reason as mentioned previously regarding conductive structures, the utilization of VAT techniques for fabricating composite sensors is rarely described in the literature. A simple example presents a fabrication of a strain sensor array from UV curable resin containing MWCNTs, adapted also for detecting humidity and temperature [626]. A more complex study deals with the integration of mechanical and electrical properties in complex 3D shapes. Here, graphene nanoplatelets in UV curable resin were printed with the SLA technique and at the same time were aligned with an electric field allowing the fabrication of 3D printed lightweight smart armor capable of damage occurrence sensing via resistance change [627].

6.5 Microelectromechanical systems

Actuators fabricated with additive manufacturing are presented even for 2D printed electronics on paper [629] so the adaptation of materials and techniques for 3D printed actuators or even microelectromechanical systems was a natural course of development. The already mentioned 3D printed loudspeaker is a good example of one of the most popular forms of electromechanical actuator, entirely fabricated with additive techniques (FDM and DIW) [417]. Another approach for the free-form fabrication of interactive loudspeakers is based on the electrostatic approach, which is simpler to build than common electromagnetic speakers [630]. The most basic and versatile working principle of actuators is a change in the geometry via thermal stress induced by electrical current flow. Using DIW of a graphene paste, a combined sensing/actuating soft, flexible structure exhibits bending with connected voltage [236]. Such an approach is quite popular in the literature, and DIW printing of ionogel was also utilized for the fabrication of soft actuators with haptic, proprioceptive and thermoceptive sensing capabilities [628]. All-printed robotic grippers with the integration of haptic and temperature sensing were done by DIW of platinum-cure silicone elastomer, and organic ionic liquid filled with fumed silica particles, allowing selective detection of different textures of grabbed objects via resistance changes measurements of ionic liquid-based sensors (Figure 6.10(f)) [628]. Other results demonstrate DIW 3D printed cellulose nanocrystal composite architectures with a high degree particle alignment [631]. Such structures inspired by nature exhibit programmable reinforcement along prescribed directions, and tailored responses to the applied mechanical load. Spatial variation in the distribution and orientation of nanomaterials can also impact the shape-changing abilities of 3D printed actuators. The orientation of alumina platelets allowed the production of shape-changing structures with programmable shape deformation (Figure 6.11(c)) [632]. An interesting approach to fabricating a micro-electromechanical relay switching device was presented, incorporating inkjet printed metal nanoparticle-based inks on three-dimensional cantilever structure, exhibiting very low on-state resistance (10 Ω) and very low off-state leakage (Figure 6.11(a)) [633]. A thermally active structure for the fabrication of actuators was also prepared from the mixture of silicone and wax materials, to fabricate electromechanical machines [634, 635]. What is interesting such electromechanical and bulk nonvolatile systems are evaluated as a building blocks for neuromorphic hardware systems [636]. Several groups have tackled the VAT polymerization approach. For example, ferromagnetic iron nanoparticles confined by the applied magnetic field within SLA printed structure exhibited regional differences in magnetic remanence, useful for teleoperation, rotation, translation and deformation [637]. A similar approach with SLA printing was used to fabricate an actuator triggered by temperature, demonstrating a simple on–off switch for electrical circuits (Figure 6.11(d)) [638]. With the use of two-photon polymerization, a magnetic polymer composite with Fe_3O_4 nanoparticles was used to fabricate an actuator for swimming microrobots [639].

3D printing of micromechanical fine structures and the ability to integrate them with electronic circuits opens the possibility to fabricate lab on chip microsystems, fostering the research toward miniaturization of biological or chemical processes, with microfluidics and miniaturized sensors (biochips). Silicon-based technology and other forming techniques for such structures typically utilize time-consuming, multistep lithographic processes to fabricate such MEMS systems [640]. An entirely DIW printed lab on a chip system contained carbon-TPU composite strain gauges, and silver-nanoparticle printed electrical leads and contact pads, deposited in multilayer process on various support layers. This biochip micro-architecture was dedicated to the growth of cardiac tissues over the course of a month and constant monitoring of its biological parameters (Figure 6.11(b)) [641]. Scaffolds for the growth and the metabolic activity monitoring of microalgae and mesenchymal stem cells were printed with an integrated sensor based on nanoparticle luminescent indicator ink, utilizing DIW technique [642]. Configurable 3D printed microfluidic lab-on-a-chip reactor for chemical syntheses of gold nanoparticles was fabricated in just a few hours with FDM technique [643]. FDM was also used for the fabrication of microfluidic polymer components with integrated nano/microparticles as biosensors to detect the AFP antigen [644]. While the SLA technique allows for the fabrication of high-quality fine structures microfluidic devices were printed this way with magnetic nanoparticles as an active approach to remove and sort E. coli bacteria from a solution [645]. The most precise 3D printing technique, also VAT resin-based, two-photon polymerization (2PP), was used to fabricate bio-inspired venous-like valves in a microscopic cage for motile bacteria trapping to improve the sensitivity of additional

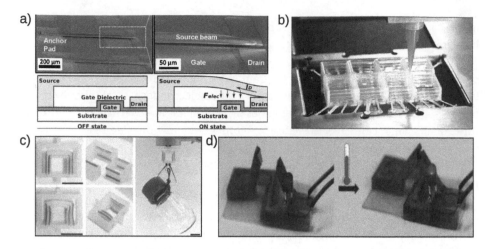

Figure 6.11: (a) 3D printed MEM switch fabricated using nanoparticle ink [633]. Reprinted with permission from American Chemical Society, copyright 2013. (b) Instrumented cardiac microphysiological devices fabricated via multimaterial 3D printing [641]. Reprinted with permission from Springer Nature, copyright 2017. (c) Magnetically assisted 3D printed composite actuators [632]. Reprinted from Springer Nature under Creative Commons CC BY licence, copyright 2015. (d) 3D printed shape memory-based electrical devices from carbon nanotubes with a temperature activated sensor in its "off" and "on" state [638]. Reprinted with permission from John Wiley and Sons, copyright 2016.

graphene-based biosensor [646]. A hybrid approach utilizing several 2D and 3D printing techniques was also used for the fabrication of instrumented cardiac microphysiological devices with piezoresistive, high-conductance and biocompatible soft materials inks, enabling the integration of soft sensors within cardiac tissue microarchitectures, further applied also to study drug responses [641].

7 Challenges

Several challenges are related to the fabrication of 3D printed electronics elements and devices, and such a process requires consideration of various aspects influencing each other, including functional materials, printing process parameters, additional postprocessing and a new approach to the components and circuit design. Although there are plenty of material types including conductors, semiconductors and dielectrics, not all of them are suitable for structural printing of electronics, thus they are applicable for planar-printed electronics. Not all materials are suitable for different deposition and post-processing parameters. Materials unsuitable for bulk elements could result in uncontrolled layer thickness posing a challenge for spatial object fabrication, which could also affect the production yield of structural electronics. While this will primarily affect the production and scalability of the components, it will also have a major impact on the efficiency and final application areas of the 3D-printed electronic components. Short-circuit, poor conductivity, nonuniform layer thickness and quality of printed paths and elements, or any other defects will negatively affect the efficiency and reliability of such electronic systems. Lastly, the integration of discrete components, and efficient fabrication of interconnects is also a challenging approach. Such technological obstacles, if addressed properly, will be the key elements for the widespread adaptation of 3D structural electronics in daily life applications. Of course, we will benefit from the advances in structural electronics if we adopt design practices to meet the requirements and capabilities of such an approach with effective design algorithms, different from today's used planar PCB design CAE tools.

7.1 Efficiency

One of the most important objectives for modern electronics is the high operational efficiency of the systems. This can be attributed to the high speed of processing, high energy efficiency, low signal-to-noise ratio, large operational memory, etc. At this stage of the development of 3D-printed electronics, one of the most challenging and the most trivial parameters is high electrical conductivity. Trivial, because for most of today's electronics, this is rarely contemplated factor, while almost all conductive components of the circuits and elements (paths, electrodes, connectors, etc.) are fabricated from the bulk metal, mostly copper or aluminium, achieving the bulk conductivity of these materials without any additional modifications or postprocessing. While most of the 3D printed conductive elements are fabricated from composites, or composite inks subjected to additional post-processing, such materials exhibit limited efficiency regarding electrical conductivity or current load. A classic example of an efficiency limitation, known from printed electronics, and also attributed to structural electronics is the performance of transistors manifested in the mentioned carrier mobility. Although a dense array of carbon nanotubes (popular for the fabrication of printed TFTs) synthesized using the

https://doi.org/10.1515/9783110793604-007

CVD technique allows a carrier mobility value of 1000 $cm^2V^{-1}s^{-1}$, close to the monocrystalline silicon and outperforming amorphous silicon TFTs. This value falls by few orders for composite printed structures. The answer is to use purified nanotubes and introduce higher homogeneity of nanotube inks or adaptation of longest possible CNTs, ultimately exhibiting much higher carrier mobility than network of short nanotubes, eliminating losses associated with contact resistance between nanotubes [647, 648]. The performance of circuits is also related to the interaction of the substrate and the printed structures. Commonly used polymers for 3D printed substrates need to be more extensively evaluated toward the negative impact on circuit performance. The same applies to the polymers used as a matrix in composites and surfactants improving the dispersion, and other chemical additives influencing the performance of composited electronic circuits [649].

7.2 Miniaturization

Inherently connected with the higher efficiency of electronics is the miniaturization of components. This is true for the conventional electronics based on PCBs and ICs that do not necessarily be attributed to the printed and structural electronics in general approach. Due to the integration into macroscale structures, sometimes the size of a table, car or building, miniaturization is not always a key factor regarding conductive circuits, antennas, sensors, resistors, capacitors or energy devices. That changes dramatically if we try to integrate printed active elements or entire integrated circuits, such as transistors or microprocessors. While still, the geometry of the entire active structure is not subjected to many restrictions; the building blocks are subjected to the physical restrictions of the resistivity, carrier mobility, etc. and here for high-performance electronics the miniaturization of 3D printable structures is required [650]. 3D printing technology has many technical limitations that was mentioned previously regarding the alignment of the nozzles determining the dimensions of printed structures. Therefore, we also find studies toward overcoming these limitations for high-resolution 3D printing.

One of the examples is the optimization for achieving circuits operating at high frequencies, related directly to the ability to produce transistors with a short channel length. Typically, for most printing and deposition techniques, a resolution of 10 μm is a technological barrier. Knowing such limitations more than a decade ago, work was begun to print transistors with a submicrometer channel length [651], and more recent reports demonstrate that 250 nm resolution can be achieved [652]. Other reports present an adaptation of sort of a nano-DIW printing with a glass nanopipette nozzle with 630 ± 70 nm diameter used to print 3D nanopillar structures with quantum dots for RGB displays [543]. An increase in the brightness of the pixel was achieved with the controlled pillar height without changing the spatial resolution of the pixel. A modified inkjet printing technique with an external electrical field was used to fabricate a high aspect ratio wall with submicron dimensions [653]. A similar approach with electro-hydrodynamic printing allowed to stack up the printed layers vertically, achieving silver or gold metallic walls with a line-width from 80 to 500 nm and a height from 200 nm to 1.5 μm [654]. Such structures could be incorporated as a 3D channel or electrodes of the transistor. Many other reports demonstrate a submicron scale deposition of materials forming as pillars, walls and bridge-like structures from metals (silver, copper, cobalt) with high-resolution electro-hydrodynamic inkjet printing [523] or electrostatic assisted aerosol jet printing of palladium nanopillar array with 300 nm 3D metal structures [503]. Even a basic adaptation of inkjet printing allowed to fabricate metal pillars from silver nanoparticle inks with a high aspect ratio, deposited layer by layer with the maximum height of a pillar reaching 10 mm, and aspect ratio 50:1, later utilized as vertical interconnects [655]. Although the resolution of 3D printing techniques has reached the scale of submicrons, it should not be compared directly with state-of-the-art electronics based on photolithography, yet it is expected that the 3D printing resolution will be enhanced further.

7.3 Connections and embedded components

Another factor to consider for the effective fabrication of high-efficiency 3D printed electronics is the integration of discrete components and packaging, which can be a major bottleneck for high-scale fabrication. Despite many benefits of introducing structural electronics, the major drawbacks are still mentioned—inferior electrical performance and lower integration degree compared to conventional solid-state electronics. One of the options to meet the requirements for high electrical performance, conventional electronics components need to be combined with additive manufactured elements. Such a hybrid approach is well known from other techniques such as printed electronics, and it is also a building block of the PCB integrating discrete components on conductive circuitry. The interest toward the hybrid approach is due to its ability to introduce freeform and new formats of applications, while still providing high electrical performance with solid-state electronics discrete components. This requires the efficient fabrication of another very important element of electrical circuits, components connections and interconnects. In electronic circuits, component leads are connected to conductive paths (copper in PCBs). Such connections both realize electrical connections and mechanically confide components in place [656]. In conventional electronics, specific materials such as solder alloys or conductive adhesives are used to make connections. In structural electronics, it is possible to use the same materials that are used to fabricate conductive paths. For instance, one of the most popularly used, silver nanopowder inks and pastes, achieves resistance values comparable to those exhibited by conventional joints [657, 658]. The use of composite material leads to higher resistance, but this might be sufficient for most applications. In the 3D structural approach, the electronic components are very often completely surrounded by the construction material, so mechanical strength is not crucial for a connection. Regardless of the materials used, an integrated system for the 3D printing of electronics and integrating discrete components would have to consist of several blocks, including 3D printing of the substrate, multimaterial deposition of conductive paths and other functional elements, precision component placement, connection fabrication (if not mounted on the deposited material of the paths) and additional packaging and sealing, with an inspection system also for quality evaluation (Figure 7.1(a)).

The hybrid 3D printing of electronics with embedded discrete components will probably create an unusual environment for the materials and mounted components, with operating conditions that they are not designed for (limited heat dissipation, thermomechanical stresses, etc.) and known from other embedded systems. What can negatively affect the reliability of such a device [659–662]? Here, the 3D deposition of materials with already mounted components might be beneficial for thermal management. Additional material can be deposited to form a heatsink printed and cured directly onto a discrete component attached to a circuit, eliminating the use of additional thermal interface materials and heatsink assembly and overall reducing the number of thermal interfaces. Nevertheless, temperature fluctuations in such embedded systems will induce thermal expansion and stress at the interfaces of materials, effectively ending with fatigue and failure. Evaluation of the failure mechanism is very important for conventional electronic systems, especially regarding the reliability of interconnects that are responsible for almost 40 % of electronic circuit failures [663]. In general, fatigue is defined as the materials' response to repeatedly applied loadings, resulting in cyclic deformation and failures. Fatigue failure starts as microstructural change, forming microscopic cracks that propagate under continuous cyclic loadings until the object structure is damaged [664]. Environment factors such as humidity and temperature causing material degradation play an important role here but internal features, microstructural defects, grain size or structure also might promote fatigue. Unoptimized printing could introduce variations or gaps in printed structures (both substrate and conductive paths) causing issues in terms of fatigue and reliability.

As presented in Figure 7.1, the general concept of additive manufacturing of hybrid circuits consists of several stages of load-bearing formation, materials deposition, component mounting and packaging. Such examples of embedded components are difficult to present, and here the most important part is the evaluation of possible approaches to the connections fabrication. Therefore, examples mentioned here cover fabrication of connections for mounting discrete components on polymer substrates, adaptable for 3D printed electronics. The most direct approach adapted from PCB technology is the use of solder alloys for mounting components on FDM-printed conductive composites [665]. This poses a challenge, while the alloy solders are tailored for copper substrates and not for composites containing polymers, thus the problems with wettability and adhesion arise, but also a thermal melting of the composite during the soldering process. Another attempt incorporates a DIW-printed paste for thermal curing or selective sintering, minimizing the negative thermal influence of the process on the polymer substrate [447]. For instance, DIW was used for the fabrication of a wire-bond-like connection with the mi-

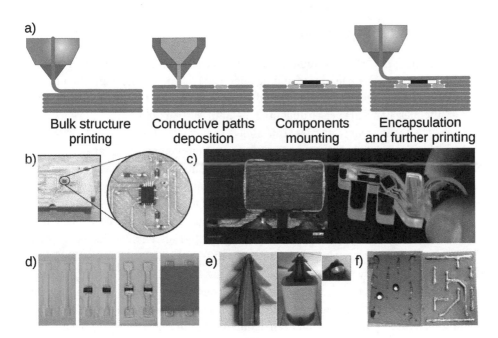

Figure 7.1: (a) A schematic diagram of the process steps in the additive manufacturing of paths, joints and connections with encapsulation. (b) Silver paste composition used for connections in DIW printed Cubesat system [425]. Reprinted with permission from Solid Freeform Fabrication Symposium. (c) Cross-sectional image of the integrated passive SMD component connected to the conductive tracks, embedded in the 3D printed sample [662]. Reprinted with permission from Taylor and Francis, copyright 2018. (d) Fabrication steps of exemplary structural electronics connection (substrate printing, components mounting, paths printing and encapsulation) [312]. Reprinted from MDPI, Basel, Switzerland, under Creative Commons Attribution 4.0 license, copyright 2022. (e) 3D-printed demonstrator with soldered LED on PS/Cu conductive substrate using a hot iron and SnPb soldering alloy [665]. Reprinted from MDPI, Basel, Switzerland, under Creative Commons Attribution 4.0 license, copyright 2021. (f) 3D printed multivibrator circuit fabricated with fusible alloy paths and connections (SnAgCu) [182]. Reprinted from MDPI, Basel, Switzerland, under Creative Commons Attribution 4.0 license, copyright 2022.

cropipette and selective metal structure forming through electrodeposition, allowing to achieve wire bonding of Cu wires, with a diameter of 800 nm [666]. DIW was also used for freeform printing of silver microelectrodes with a minimum width of 2 µm, used as an interconnect for the gallium arsenide-based SMD LED array [667]. Direct printing of liquid metal allowed to achieve lines with a minimum width of 1.9 µm [76] used for the connection of LEDs, and later employed for the interconnects in an array of microLEDs. Also, a modified FDM technique for printing molten metals (FDMm) allows for fabrication at the same time paths and connections, eliminating the need for additional processes [182]. This is a good example that unlike in conventional methods for forming interconnections, freestanding 3D interconnects can form a single path with much fewer (or even none) intermediate connections, which in regular PCBs, are adding additional contact resistance and often lowers the reliability of the whole circuit.

7.4 Multimaterial printing

For the selective deposition techniques such as FDM, DIW, inkjet or AJP, the multimaterial printing is mostly a matter of adding additional nozzle for the deposition of new material (i. e., polymer substrate and conductive paths). Most of these techniques are also easily interchangeable regarding the kinematic system, while they all rely on the XYZ set with exchangeable nozzles. This allows to fabricate FDM substrate with DIW paths and inkjet printed transistors or other active components, as an example. Some alternative attempts utilize a single nozzle for multimaterial printing, such as filament retraction and loading of a new material already commercially available for FDM printers. An interesting approach was also introduced for multimaterial DIW printing with a single nozzle. A rapid switching between different materials (up to 50 Hz) in an array of printing nozzles capable of depositing up to eight materials, enabled the formation of soft-robotic walker's legs were printed with two different silicone rubbers of different stiffnesses [668]. However, multimaterial 3D printing with techniques utilizing selective curing or binding of materials, instead of selective deposition has been challenging due to the difficulties of exchanging a liquid material in a vat or powder on the bed. The "one material at a time" is a fundamental restriction for SLA or SLS techniques. That limitation is often circumvented by separate printing processes of polymer substrate and depositing layers from different materials [669, 670]. Such an approach significantly slows down the process and increases the overall manufacturing time. Other attempts introduce various techniques for material replacement, usually providing separate vats with resins or hoppers with different powders, and the process repeats for each layer, fabricating mostly multilayered structures instead of truly multimaterial [671–675]. Besides introducing additional steps to the printing procedure that significantly slows down the whole process, they also pose a risk of cross-contamination between materials, more important with frequent material change for hundreds of layers.

The two most promising approaches for resin and powder-based techniques regarding truly multimaterial printing are based on prompt material exchange and do not introduce many modifications to the solidification process itself. The recent example of the VAT polymerization technique is based on a rapid multimaterial projection micro-stereolithography using dynamic fluidic control of multiple liquid photopolymers within an integrated fluidic cell [676]. With this approach, more than 95 % of the material inside the vat is exchanged within a few seconds, with a limited degree of cross-contamination between different materials, which is comparable with regular hold time between layers in regular SLA technique. Highly complex multimaterial 3D microstructures were fabricated with this active material exchange process, also a multiparticle-loaded structure with embedded metal (copper) and ceramic (alumina) nanoparticles, demonstrating the fabrication of a thermoresponsive and electroactive hydrogel bilayer beams. Another attempt of powder-bed technique application uses a similar approach of powder exchange within a chamber using a vacuum extraction of powder and deposition of a fresh layer [677].

Also, these techniques have their limitations regarding the adaptation of UV curable resins (applied to Multijet techniques, but also VAT polymerization). Due to the mentioned restrictions related to the fabrication of highly conductive UV-cured composites, only dielectric, photonic, magnetic or highly resistive materials can be obtained, limiting severely application of these techniques. Already many problems with the adaptation of VAT techniques for 3D printed electronics are observed such as the negative impact on the polymerization process with the high concentration of metal or ceramics and oxides nanoparticles [559, 560, 678], the need for additional sintering procedures to obtain conductive structures with silver NP for ink-based techniques [679, 680]. Examples of multimaterial printing for VAT techniques and for voxelated deposition are presented in Figure 7.2.

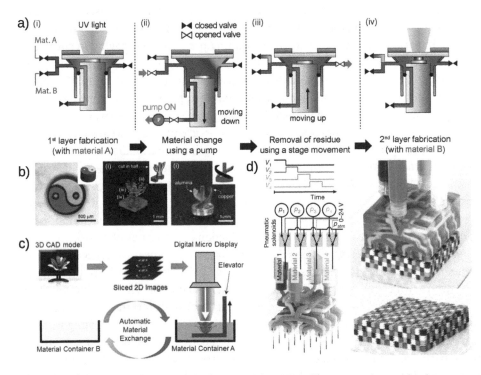

Figure 7.2: (a) Schematic illustration of the MultiMaterial-PμSL overall process and material exchange process, and (b) multimaterial micro 3D structures fabricated with MM-PμSL containing two or three different materials, also resins loaded with copper and alumina nanoparticles [676]. Reprinted with permission from Elsevier, copyright 2019. (c) A workflow illustrating the process of fabricating a multimaterial structure based on PμSL and exchange of VATs [672]. Reprinted from Springer Nature, under Creative Commons CC BY License, copyright 2016. (d) Schematic of multimaterial 3D printhead operation, with voltage waveforms controlling the extrusion pressures, and voxelated matter produced by MM3D printing using a 4 × 4-nozzle, four-material printhead [668]. Reprinted with permission from Springer Nature, copyright 2019.

7.5 Scalability

Another important element in the puzzle of the mass introduction of 3D printed electronics, besides efficiency, miniaturization or multimaterial printing is the ultimate integration of all process components in the most effective way to enable scalability for mass production and mass customization, which is expected from additive manufacturing. The fundamental challenges of 3D structural electronics devices today emerge from the ad hoc combination of manufacturing methods and materials, for which they are not initially designed for [662, 681]. The dependencies and interactions between various fabrication processes and large groups of materials and components are not addressed these days, thus the characterization level of the technology is low, impairing the potential for effective exploitation. One of the most obvious reasons for the optimization of production is cost reduction, allowing direct competition with PCB-based electronics. But this is not so obvious where to start with cost optimization when the technology is in its infancy stage. The majority of 3D printing processes are lagging in speed and efficiency compared to conventional production therefore slow printing speed creates an obstacle to their application in industries that require mass production. This might not be a major problem for the mass customization approach, with already prefabricated systems modified with additional 3D printing processes. But if we want to fabricate fully integrated 3D electronics systems with complex architectures we need to improve the speed of 3D printing to maintain production efficiency at a similar level to the conventional process. And the improvement in speed is not only related to the deposition techniques. Almost all 3D-printed components require post-processing of some kind, to remove supports, improve their mechanical properties or accuracy or finally provide an aesthetic view. We already know that for rapid prototyping it is efficient to use human labor, but when it comes to the mass production of thousands of 3D printed parts, automatization of cleaning, grinding and post-processing is a necessity. A good example is mature powder bed techniques such as SLM or EBM, already well introduced in the industry, for which the printed part after removal from the plate, needs additional heat treatment to eliminate accumulated residual heat that can distort the part during cooling. We also have examples from the work on 3D printed electronics, dealing with the problem of a suitable curing process that simultaneously will be fast and not damage (even partially) the low-temperature polymer substrate [447, 682]. For 3D printed electronics, we do not have any guidelines in that matter, but building on the experience of more matured additive techniques we should already be prepared for the whole workflow optimization, and not just focus on the materials, printing parameters or efficiency of the final components, as it might be not enough to successfully introduce 3D printed electronic to the market.

7.6 Designing process

Lastly, having all the needed functional materials, optimizing all printing parameters to the maximum, and introducing the most efficient 3D printing techniques dedicated purely to electronics, we need to make a full circle and ask the primary question: how to design a 3D printed electronic circuit? Fundamentally, such a design will be universal for all techniques and types of materials. The ability to design truly 3D printed electronics with the set of essential design rules exploiting all the advantages over the conventional PCBs might be very challenging and might pose a most distinctive bottleneck [683]. While today's electronic design automation (EDA) software for PCB design is capable of designing multilayered circuits, with an almost infinite number of layers, and different board outlines, this is still a 2D approach for the fabrication that might be efficient enough for the design of a microcontroller embedded in a flat table, but not an advanced circuit inside the aircraft part or well-fitted wearables. Moreover, such design software should also take into account process limitations. Conformal or 3D printing is more complex than printing on flat substrates, with more design and fabrication considerations. Besides the topology of the circuit, the accessibility of the print head to the surface of the 3D structures needs to be considered and that requires toolpath planning with the alignment of the print head to 3D parts [684]. Already some attempts are introduced to solve these problems and research on software that facilitates 3D modeling of electronics is being progressed such as topology-aware routing of electric wires in FDM-printed objects [683], algorithms for selective laser sintering of spatial DIW printed objects [459] or guidelines and process planning considerations for layered manufacturing with embedded components [685]. One step further goes a DIW system equipped with the closed-loop tracking system of the motion on the target surfaces, not only introducing the conformal printing of the circuit, but allowing 3D printing on dynamic surfaces such as moisture sensors printed during the movement of the hand, or deformation sensors were conformally printed on the surface of the porcine lung [609, 686].

8 Perspectives for additive manufacturing of electronics

While the previous chapter was focused on challenges that 3D printed electronics is facing in its current stage of development, this chapter presents new directions fostering progress in research areas of new techniques and materials, but also new products and systems emerging in the market. Current progress in additive manufacturing is so rapid that even during the last months of the preparation of this book new techniques and applications emerged. Moreover, only a limited group of the most popular 3D printing techniques and materials was presented here, and there are tens or hundreds of other techniques and hybrid approaches suitable for structural electronics fabrication. Work on improving the parameters of 3D printed circuits with new materials and modifications will be presented. Interdisciplinary approaches combining electronics and biomedicine will also be covered in this chapter, not to mention more exotic scopes of self-healing materials or printing below freezing point. Finally, three-dimensional circuits will go one step further to "four-dimensional" structured electronics produced by additive techniques (4D printing).

8.1 Market development

As mentioned previously, the expectations toward the 3D printed electronics market are to grow fast within the next decade. As for all industrial applications, besides the development of new materials, techniques and identified applications, this will require an appropriate technological background supplied to the market in the form of off-the-shelf production solutions. Thus, today mass produced 3D printed electronics sounds like a distant sci-fi, a number of companies are already working on the practical applications, foremost on the development of equipment tailored for such an approach to electronics manufacturing. One of the pioneers of structured electronics was Eric MacDonald. He created a fabrication system consisting of two 3D printers, a DIW station and a CNC machine tool operated by a 6-axis robotic arm [426]. He also reported and commercial application of a functional 3D printed electronics device as a CubeSat satellite module [424]. A similar concept was developed by the team from Harvard University, US, resulting in the spin-off Voxel8 with a complete system, commercialized as the world's first general-purpose 3D electronics printer on the market capable of producing a fully spatial electronic circuit. This device combined an FDM head with a DIW module. At the heart of this development is particle-free, low-viscosity reactive silver ink based on $AgC_2H_3O_2$, curable at 90 °C, exhibiting an electrical conductivity of $>10^4$ S/cm [420]. The company demonstrated printed electromagnets and functional devices such as a quadrocopter. The discrete electronic components need to be placed manually with the printing process paused. Eventually, after selling the company, Voxel8 changed its area of operation

https://doi.org/10.1515/9783110793604-008

and the printer is no longer available on the market. A similar DIW approach was presented by Voltera with a V-One Circuit Board Printer [461]. Using a conductive as well as an insulating ink, the V-One allows printing of two-layer circuit boards. The growing demand for devices to produce fully printed structural electronics has been recognized by some companies. Over the past few years, they have decided to develop complex equipment, consisting of multiple modules for the application and curing of conductive materials, as well as for the additive printing of substrates. In addition, modules for the automatic placement of electronic components are also very often presented. One of the first companies to offer professional equipment for the production of 3D electronics is nScrypt, established during the MICE program, now developed a series of nScrypt 3Dn equipment with microdispensing DIW, FDM, micromilling and drilling, pick and place for discrete components, laser processing, sintering, heating, UV and photonic curing, plasma pen or even real-time process view [687]. Other companies using hybrid DIW approach include M4 3D Printer [688] or Neotech AMT 15x BT [689]. A DIW approach is one of the most popular while it is well established on the electronics market as a technique for applying solder pastes, adhesives, sealants or other high viscosity materials, and this technique is highly explored for the bioprinters market by companies such as Cellink, Allevi, BioAssemblyBot or RegenHu. A bit different approach to structural electronics is Aerosol Jet Print from Optomec, US [470]. The offered platform does not create a structural part, but it is used for the conformal coating on an existing 3D substrate. Along with Neotech GmbH, Germany, Optomec developed a system equipped with Light Beam Sintering (LBS), a noncontact, photonic sintering technique for low-temperature substrates such as polymers. This approach is very popular for the production of conformal antennas used in mobile devices, including LTE, NFC, GPS, Wifi, WLAN and BT with measured performance comparable to other production methods. Another ink-based approach is developed by Nano Dimension with DragonFly 3D printer for additive manufacturing of multilayer PCB via multijet printing of both dielectric and conductive materials [469]. It produces high electrical conductivity wires (5.73×10^{-8} Ωm, around 30 % of copper) and can print vertical vias, but it is mostly restricted to thin 2.5D PCBs. As mentioned here, systems do not represent ready-made technology, as they are only specialized tools. Manufacturers do not offer support in the selection of materials and their technological parameters, which makes it quite difficult to determine the final results. A variation of printing and processing modules (DIW, AJP, inkjet, FDM, UV curing, photonic, etc.) is preferred for R&D work, but at the same time, it is difficult to deploy large-scale mass production of 3D structural electronics.

8.2 Core-shell nanomaterials

Electrically conductive nanoparticle inks, pastes and composites are dispersion of metallic nanoparticles, widely used as liquid medium inks for the fabrication of electrically conductive patterns in 3D printed electronics. While with nanomaterials, we observe an extensive agglomeration related to the strong van der Waals forces interaction between nanoparticles, causing poor dispersion or nozzle clogging, in most cases each metallic nanoparticle is encapsulated with organic additives to avoid agglomerations. The same organic additives, however, prevent contact between metallic nanoparticles limiting the electron transport and current flow therefore initially electrically nonconductive nanoparticle inks require sintering processes to decompose the organic additives. Organic additives are also to prevent the extensive reactivity of the nanoparticles, mainly oxidation and, therefore, are one of the key components of nanoparticle inks. Single-element metallic nanoparticles (Ag, Au, Cu) are widely used for formulating inks and pastes, with silver nanoparticles being highly preferred for fabricating conductive patterns due to the highest electrical conductivity from all metals and its excellent oxidation stability. Alternatively, gold nanoparticles also have good electrical conductivity and oxidation stability, but their price tag is much higher due to the cost of gold on the market. Therefore, the high interest is focused on copper inks, with the Cu cost much lower than silver and gold and comparable electrical conductivity with silver. However, copper nanoparticle inks as a substitute for expensive silver nanoparticle inks may not be so easy to implement, because copper nanoparticles are highly reactive under ambient atmosphere and tend to oxidize easily into copper oxides (semiconductors or dielectrics) during sintering at high temperature [690, 691]. Here, a new group of bimetallic core-shell nanoparticle inks emerges, still in the research phase, trying to step into the market. The core-shell nanoparticles comprise two different materials, where the outer shell is usually oxidation stable (Ag, Au, Ni) preventing the inner core (i. e., Cu, Al) from oxidation and corrosion [692, 693]. Besides the limited environmental influence on the main core material, core-shell nanoparticles offer several advantages in achieving the desired electrical, optical or chemical by tuning the core-to-shell materials ratio. Also, the economic aspect plays here an important role, while it is easy to reduce the load of expensive materials. The copper-silver (CuNP@Ag) core-shell bimetallic nanoparticles are one example, where a thin silver shell protects the copper core from oxidation [692, 694]. These nanoparticles are gaining more traction in the fabrication of electronic inks and composites while the load of expensive silver is significantly reduced, retaining exceptional electrical properties [695]. Other core-shell bimetallic nanoparticle inks are also developed such as Ni-coated [696] or Sn-coated [697].

8.3 Controlled organization and patterning

Specific properties of 3D printed structures can be obtained by the selection of materials, adapting techniques, optimization of printing and curing parameters, and finally by the design of the element. Another less usual option is the modification of materials' properties during the printing process. This can result in the controlled formation of microstructures or spatial organization of filler phase in the 3D printed materials, fostering an increase of functionality or the final performance of 3D printed electronic elements and devices. Controlled organization and microstructure formation can be realized at the macro, micro and nanoscale. One approach is inducing anisotropy via the organization and formation of particles during the printing of the materials, mostly implemented for the DIW and FDM extrusion techniques. Here, shear forces in the nozzle enable the alignment of nanoparticles [236, 698, 699]. A controlled evaporation and cooling at the nozzle also promotes the self-assembly of nanoparticles during the printing (Figure 8.1(b)) [97]. Particles dispersed in the medium (e. g., nanofibers, nanowires or nanoplatelets) experience shear stress and realign their principal axis with the direction of the laminar flow inside the nozzle. Such phenomena can be used to program anisotropic stress–strain behavior in 3D printed structures and to program shape deformation in response to external stimuli (Figure 8.1(a)) [700–702]. Fluid shear patterning allows also to print self-supporting structures by nanoparticle influence imparting stress thinning [703]. Also, the orientation of particles can be beneficial where a low concentration of filler is needed to prevent nozzle clogging, but the final mechanical stiffness should be higher than for randomly arranged filler particles [631]. Occasionally, higher share rates at printing allow modulation of electrical properties, enhancing conductivity for composites with oriented nanoparticles (i. e., graphene or nanotubes) [99]. This approach was also used for the fabrication of lithium-ion battery electrodes, where nanoparticles (GO) were aligned along the extruded direction improving the electrical conductivity of the entire electrode [704]. Also, piezoelectrical properties can be enhanced by inducing oriented semicrystalline structures in ceramic nanocrystal-polymer hybrid materials, and with orientation adjusted to the direction of applied stress [705].

As an analogy to the operation of LCD displays, fluid shear patterning can also program optical functional properties, by orienting nanoparticles and inducing the optical response to polarized light [563]. After the deposition, the most often used self-assembly phenomenon is solvent evaporation in the inkjet technique, developing capillary forces promote the self-assembly of the printed particles (i. e., graphene platelets or nanotubes) and organizing them in a planar, anisotropic structure in the simple drying process [706]. Patterning and assembly via solvent evaporation also allows generating the desired microstructure with DIW or ink-based techniques, such as various freestanding 3D structures [707], nanoarches [97] nanopillars and dome structures [708, 709]. While for deposition techniques a shear and evaporative assembly and patterning are dominant, they are difficult to implement in the VAT applications. Here, one of the approaches is acoustic patterning, also referred to as acoustic tweezers, used to move or concentrate the particles using the interaction between acoustic waves and the surrounding solid, liquid or gaseous media [710, 711]. Such patterning was used in VAT techniques for 3D printing of patterned nanoparticle-polymer composites in the shape curves and crisscross patterns and to create parallel lines as a conductive microstructures (Figure 8.1(d)) [712–714]. Electrical field patterning for aligning electrically conductive particles under unidirectional electric field has been used to fabricate patterned and organized structures with VAT resin based techniques. Here, graphene platelets and carbon nanotubes were printed with field-induced particle orientation to enhance mechanical properties

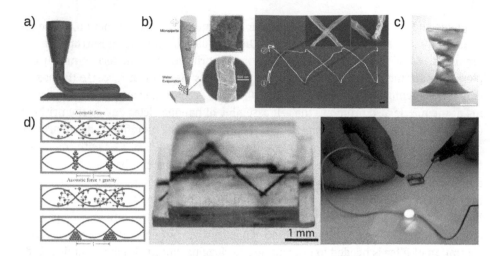

Figure 8.1: (a) Fluid shear force-induced alignment of nanoparticles during direct ink writing 3D printing [702]. Reprinted with permission from Springer Nature, copyright 2016. (b) Schematic diagram of GO nanowire fabrication by pulling a micropipette filled with an aqueous GO suspension, and suspended rGO junction nanostructure assembled using two concatenated rGO nanoarches [97]. Reprinted with permission from John Wiley and Sons, copyright 2014. (c) Multimaterial magnetically assisted 3D printed composite structures with SLA technique [632]. Reprinted from Springer Nature, under Creative Commons CC BY License, copyright 2015. (d) Acoustic patterning for 3D embedded electrically conductive wire in stereolithography, with embedded zig-zag stitch wire pattern of electrically conductive copper nanoparticles [714]. Reprinted with permission from IOP Publishing, copyright 2017.

forcing controlled crack propagation or selectively aligning particles and cure regions of the print to construct a nacre-like structure of aligned graphene nanoplatelets not only enhancing the strength of the composite by resisting crack propagation but also introducing electrical properties, enabling to sense damage due to a change in resistance [627, 715–717]. Magnetic field can also be incorporated in similar manner as electric field for VAT techniques. Magnetic patterning here is the alignment and distribution of magnetically active particles (ferromagnetic) with external, directional magnetic fields modulating magnetic properties of printed structures or tuning mechanical properties via control over the orientation and location of reinforcing nanomaterials [138]. Using an SLA printer with five orthogonal electromagnets and UV-curable resin filled with magnetic nanomaterials changed the residual magnetic flux density in composites [718]. Also, a programable magnetism was used for remote actuation with external magnetic fields, executed on the cantilever beam with vertical fiber alignment and tailoring the local concentration of magnetic nanomaterials (Figure 8.1(c)). Such an approach also allowed the fabrication of complex structures such as a magnetically actuated impeller or wheel [637].

8.4 Bioelectronics

Additive manufacturing has an established position in biomedical engineering, especially in the production of tailored implants. Also, printing techniques are highly explored in the field of biomedical sensors or 3D printed lab on chip microsystems also mentioned in previous sections. The obvious step forward is the combination of both 3D printed electronic devices and implants for biomedicine applications [719–721]. One interesting aspect is the enhancement of functionality of biomedical applications with embedded electronic systems. The challenge here is the biocompatibility of materials used for 3D printed bioelectronics, allowing successful device implantation to the tissue [722, 723]. The great interest is in bioresorbable materials with no need for removal [724, 725]. Research on the biomedical applications of carbon-based nanomaterials suggests their biocompatibility, which could foster the development of 3D printed bioelectronics [726]. Different strategies with biocompatible materials are already explored with the printed electronics approach including ultralow voltage in vivo neural stimulation electrodes implanted for 6 weeks in mice, made of PEDOT:PSS hydrogel, 10 times outperforming conventional platinum electrodes [727], conformal bioelectronic system for real-time brain monitoring made of silk fibroin [728], bioresorbable sensors, transducers and wireless monitoring systems of vital pressure in brain, including biodegradable poly(l-lactic acid) piezoelectric nanofibers [729] and a flexible, biodegradable pulse sensor for measuring arterial blood flow with noncontact wireless monitoring [730]. Several attempts have already been made toward the adaptation of additive manufacturing for fabricating 3D components with biomedical applications in mind. One approach covers the use of the DIW technique to print carbon nanomaterials based electrically conductive and mechanically flexible scaffolds used for in vivo experiments for electrically stimulated cell proliferation [99]. After 30 days of experiments, the scaffold integrated with the host tissue, with new capillaries grown into the implant, without signs of inflammation or graphene accumulation in the filtration organs. One of the most impressive achievements covers the use of DIW printing of a pair of left and right bionic ears capable of enhanced auditory sensing for radio frequency reception and music listening (Figure 8.2(a, b)) [731, 732]. A regenerative scaffold with living chondrocytes in the shape of an ear was printed, along with a coprinted coiled antenna inside the ear also DIW printed from silver nanoparticle paste. A tactile sensor conformally printed on the curved substrate, here directly on the human skin, was also DIW printed from silver nanocomposite paste, providing the detection of pulses, finger motions and moisture (Figure 8.2(d)) [608, 686]. Later this system, equipped tracking system of the target surface was used to deposit strain-sensor electrodes on the surface of the porcine lung and measure the deformation by electrical impedance tomography [609]. A 3D printed motion-powered smart dental implant was also prepared for biomodulation therapy of soft tissue (Figure 8.2(f)) [733]. Nanocomposite paste with $BaTiO_3$ piezoelectric phase and conventional dental materials was DIW printed in the shape of a crown,

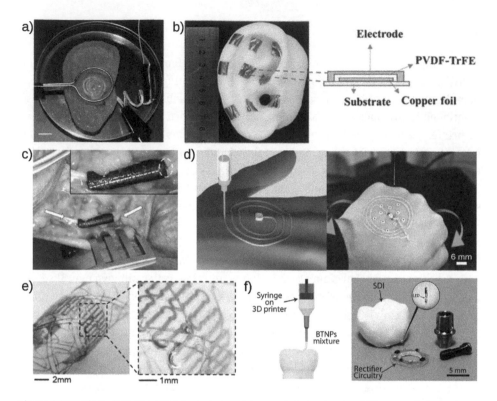

Figure 8.2: (a) Three-dimensional interweaving of biology and electronics via additive manufacturing to generate a bionic ear (biological—chondrocytes, structural—silicone and electronic—AgNP-infused silicone) [731]. Reprinted with permission from American Chemical Society, copyright 2013. (b) 3D-printed bionic ear for sound identification and localization based on a PVDF-TrFE film [732]. Reprinted with permission from John Wiley and Sons, copyright 2022. (c) Photograph of a tubular 3DG nerve conduit that was implanted into a human cadaver via longitudinal transection and wrapping around the ulnar nerve (white arrows) [99]. Reprinted with permission from American Chemical Society, copyright 2015. (d) Schematic images of direct writing of conductive ink on human hand, and demonstration of an LED connected to the a wireless power transmission system [686]. Reprinted with permission from John Wiley and Sons, copyright 2018. (e) Fully AJP printed, wireless, stretchable implantable biosystem toward batteryless, real-time monitoring of cerebral aneurysm hemodynamics [611]. Reprinted from the John Wiley and Sons under a Creative Commons CC BY License, copyright 2019. (f) 3D printed, human oral motion-powered smart dental implant for in situ ambulatory photo-biomodulation therapy [733]. Reprinted with permission from John Wiley and Sons, copyright 2020.

used as an energy harvester powering micro LEDs delivering light to treat the tissue surrounding the crown. The energy was harvested here from chewing and tooth brushing.

8.5 Biodegradable 3D printed electronics

The massive popularity of electronics prompts their rapid replacement, leading to an increasing e-waste problem. This is addressed with transient technology, an emerging field requiring materials, devices and systems to disappear with minimal or nontraceable remains after a period of operation [734]. In recent years, this trend has been extended to electronic systems and it will significantly impact the development of advanced electronics [735–737]. In additive manufacturing, eco-materials are very popular (i. e., PLA, PCL, PHA/B) matching the performance of popular polymers [738, 739]. Quick literature analysis reveals application of various biopolymers (PLA, PDLLA, PCL, PHA/B, shellac and cellulose) also for electronics [740–744] waterborne polymers and green-solvents [243, 362, 745, 746] carbon nanomaterials toward biodegradation [747, 748] and recent research on Zn and Mg particles for biodegradable electronics [744, 749–752]. Transient electronics also address a medical topic of mentioned earlier bioresorbable electronics for implants [720, 737, 753] or even exotic edible electronics [754, 755]. Most of such materials are highly adaptable to 3D printing [756–758]. The efficiency, performance or reliability of electronics with eco-materials will significantly impact the adoption of biodegradable 3D printed electronics [734]. A few examples presented in the state-of-the-art achievements on the fabrication of structural electronics deal with new, eco-friendly materials, such as lithium-ion batteries fabricated with a DIW technique from a water-based graphene oxide-based composite paste [578] and a 3D printed miniature water-based zinc-ion battery [583]. Although other components were less eco-friendly like solid electrolytes of highly concentrated graphene oxide and LFP/LTO as cathode and anode materials, such adaptation of water-based materials is a popular trend in the fabrication of electronics. Examples of transient electronics are presented in Figure 8.3.

8.6 Fusible and liquid metals

Today the most investigated group of materials are conductive thermoplastic filaments, conductive inks and pastes, used with FDM, DIW, inkjet or AJP. The primary obstacle for materials branded as "conductive filaments" or "conductive pastes" is high resistivity (a few orders of magnitude below bulk metals). On the other hand, highly conductive inks for inkjet and AJP are insufficient for fabricating bulk macrostructures. An alternative approach for low-resistivity circuits is the direct deposition of liquid metals. Metals have been used in additive manufacturing for decades in the form of powders (titanium, steel, etc.), sintered or melted with high-power lasers or electrons in high-temperature processes. On the other hand, low-temperature melting metals and alloys are also employed in additive manufacturing. Droplets of molten metal deposited with a technique similar to inkjet were elaborated to fabricate 3D structures

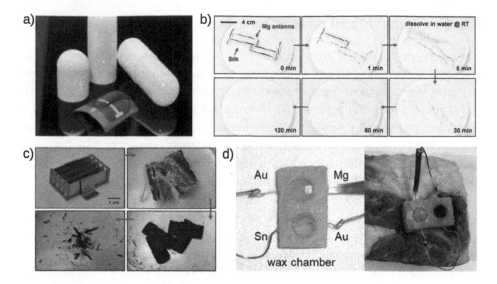

Figure 8.3: (a) An entirely "bio-OFET" utilizing a caramelized-glucose substrate, Al gate electrode with an adenine/guanine gate dielectric, and the nontoxic textile dye indanthrene yellow G functioning as the semiconductor, fabricated on a gelatin capsule [755]. Reprinted with permission from John Wiley and Sons, copyright 2010. (b) Transient RF system: an antenna built with Mg on a silk substrate illustrates the process of dissolution in DI water within 2 hours [751]. Reprinted with permission from John Wiley and Sons, copyright 2013. (c) Battery pack consisting of four Mg-Mo cells in series with the visualization of the dissolution process of the battery [752]. Reprinted with permission from John Wiley and Sons, copyright 2014. (d) Color-changing temperature sensor made of fully edible materials with the chamber made of wax placed on a piece of pork rind [754]. Reprinted from the American Chemical Society under a Creative Commons Attribution Noncommercial License CC BY 4.0, copyright 2022.

from tin-based material (Figure 8.4(a)) [181]. Solder alloys were also used in consumer-level FDM printers, due to similar extrusion temperature as polymers. Several teams presented structural elements made this way (Figure 8.4(c)) [759–762]. Therefore, the idea to use fusible metals for circuits fabricated with the FDM technique emerged. In 2009, at the University of Bath this approach was tested by providing a simple form of a FDM printed electronic circuit with a RepRap printer, deposited on flat ABS substrate [763]. Other proof of this concept was presented 4 years later, only in the form of a "circuit pattern" and not fully functional electronics circuit (Figure 8.4(d)) [764] with a later attempt to fabricate a fully functional bistable multivibrator with the FDM printing of fusible alloy paths [182]. Also, an approach for a double-layer PCB was evaluated with the use of bismuth-based alloys in the form of paste for DIW printing [765]. A bismuth-indium alloy was also used with electro-hydrodynamic printing to form vertical interconnects (VIAs) for additively fabricated two-layer circuits with a battery and LED [766]. A fully functional system demonstrating ink-jet printing of molten alloys was also demonstrated using, StarJet, capable of printing solder alloys but also molten aluminium [767].

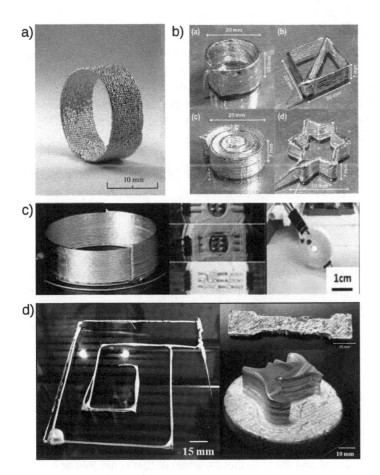

Figure 8.4: (a) Cylindrical tube fabricated by depositing molten metal tin droplets [181]. Reprinted with permission from Emerald Publishing Limited, copyright 2008. (b) Liquid metal additive manufactured gallium and its alloys structures [768]. Reprinted with permission from John Wiley and Sons, copyright 2018. (c) Highly conductive 3D printed low-melting metal alloy filament for circuitry fabrication [759]. Reprinted with permission from John Wiley and Sons, copyright 2017. (d) Fused filament fabricated 3D printed Sn60Bi40 low-melt alloys in the planar maze toolpath simulating conductive path, ASTM E8 standard tensile specimen and a souvenir logo [761]. Reprinted with permission from Springer Nature, copyright 2018.

The alternative approach uses metal alloys remaining as a liquid at room temperature, such as eutectic gallium-indium alloy, still exhibiting high electrical conductivity, easily deformable and self-healable (Figure 8.4(b)) [75, 768–771]. Interestingly, most of the liquid alloys and metals after 3D deposition in ambient atmosphere maintain their 3D structure with the formation of solid-phase oxides on the surface [772, 773]. Another approach is to print liquid metal with DIW onto a cooled substrate below the solidification point of the metal to maintain the final structure and fabricate bridge-type structures without any supporting material [455]. Liquid metals were also used to fabricate microstructures, including coils and bridges, with a minimum line width of 1.9 µm using a narrow-diameter nozzle made of a glass capillary [76]. Another approach uses 3D printing to formulate the mold with microfluidic channels (67 µm diameter) filled with liquid metal, resulting in a highly flexible coil/antenna operating in a standard near-field communication (NFC) system (13.56 MHz) [774]. Different approaches include the fabrication of composite pastes or gels with microdroplets of liquid metal that can be printed with the DIW technique, but initially is not conductive. Under the applied stress/pressure, liquid metal droplets agglomerate and form an electrically conductive network, which is an interesting approach to fabricating electromechanical sensors [775, 776].

8.7 4D printing

4D is one step further than 3D, although it is not possible to realize the geometrical constraints of our universe. Described in the literature, 4D printing is still a 3D printing (sometimes even 2D) approach incorporating geometry modifications in time [777]. The shape or physical properties change as a function of time, in response to external stimuli (temperature, chemical agent, light, etc.) [777–782]. One of the popular stimuli is light, which is an attractive stimulation method, enabling high precision via intensity and wavelength regulation. A light-responsive drug delivery device was fabricated with 3D printing, exhibiting shape change by photothermal heating activated by near-infrared (NIR) light exposure [783, 784]. With a precise 2PP technique, microsized structures with gold nanowires and a liquid crystalline matrix were printed, and rapidly changing shape upon photo stimulation [785]. A hydrogel composite paste was used to 3D print architectures with a DIW technique and shear-induced alignment that could fold and twist when placed in water [702]. Such flower-like structures exhibited anisotropic swelling behavior mimicking the flower folding when wet [702]. Also, magnetic-responsive structures were 3D printed from magnetic nanoparticle polymer composite that was changing its shape under the alternating magnetic field, inducing temperature increase [786]. Magnetic stimulation was also used for the control of soft robots printed with the DIW technique from PDMS filled with iron nanoparticles with fast actuation (under a second) [787]. Composite paste based on a PLA-based mixture with ferrite oxide was used to print structures exhibiting shape recovery under magnetic stimulation, and it was applied for the fabrication of an intravascular stent, which would shrink when magnetically heated, and expand on cooling after implantation [788]. Examples of a shape changing additively fabricated "4D printed" elements are presented in Figure 8.5.

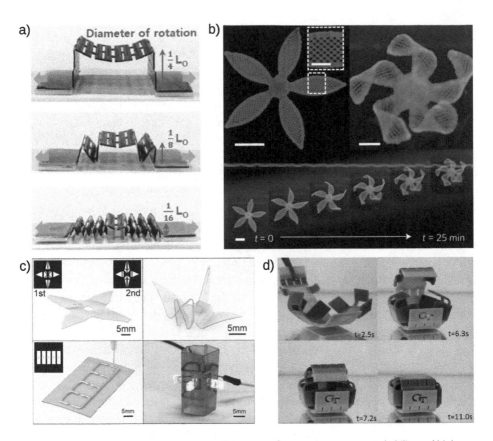

Figure 8.5: (a) 3D printed encapsulated origami electronics for extreme system stretchability and high areal coverage [780]. Reprinted with permission from American Chemical Society, copyright 2019. (b) Complex flower morphologies generated by biomimetic 4D printing, with time-lapse sequences of the flowers during the swelling process [702]. Reprinted with permission from Springer Nature, copyright 2016. (c) Desolvation induced origami of photocurable polymers fabricated with stereolitography, presenting as-cured flat pattern and desolvated origami shape (top) with an example of a box LED device folded by the rectangular sheet (bottom) [781]. Reprinted with permission from John Wiley and Sons, copyright 2016. (d) Sequential self-folding box structure fabricated from 3D printed digital shape memory polymers that are heat activated [782]. Reprinted from Springer Nature under a Creative Commons CC BY License, copyright 2015.

9 Conclusions

We live in a time of rapid development of new technologies such as nanotechnology, artificial intelligence, Industry 4.0, and mentioned here, additive manufacturing, which is expected to be a key growth driver for new developments in many fields of technology. Many of the current applications of additive manufacturing are based on the most explored techniques, such as fused deposition modeling or laser sintering and melting of metal powders. These techniques have been exploited for decades and are the most optimized for high-scale production of components. However, this approach is efficient only for selected regions of industry such as automotive, biomedicine, construction works or customization and rapid prototyping or servicing. Although making its space in many mentioned regions, additive manufacturing is still a growing market, and new approaches are developed every year focused on new materials, alternative printing methods or the development of dedicated software and design approaches. All that is to make additive manufacturing more accessible for the industry. The ultimate goal is to promote the 3D printing approach for the fabrication of final products instead of the most common these days, which are prototypes and demonstrators. The ability delivered by 3D printing allows the introduction of new design techniques, that besides new forms of components, they also allow the most anticipated advantages of energy and materials reduction, simplification of production processes that could save time and money, also promoting a lesser waste generation. The future of 3D printing lies in the simplification of the assembly stage of the production, and higher automatization of the whole production process, ultimately allowing for a higher degree of integration, which is a key factor for higher efficiency, lower weight and new applications of advanced equipment but also surrounding us with everyday products and appliances. Once these challenges are solved, the exciting new technology of 3D printing will address the most important challenges of society, and also simplify the everyday life of ordinary people.

A growing and important direction of development of additive manufacturing is the new level of fabrication of electronic devices. In order to fully exploit the potential of additive manufacturing in electronics technology, especially in commercial applications, it is necessary to adapt the existing processing techniques or elaborate new approach procedures for the use of functional electronics materials. This creates a special challenge for adapting to highly efficient mass production techniques. Initial attempts to produce conductive circuits and components on a large scale, as in the case of the conductive paths, inductors, antennas or sensors, have paved an initial way to bridge the gap between research results and practical applications in the micro and macro world. This is still progressing at a slow pace, but we need to keep in mind that current professional applications of selective laser melting of metal powders techniques took more than two decades to become an industrial-level application for the professional fabrication of highly reliable components for aerospace or medical applications. The noticeable application barrier associated with the marked difference between the properties of laboratory experiments results, often with remarkable properties, and those obtained at a

https://doi.org/10.1515/9783110793604-009

larger scale of production or higher level of technology readiness, such as in the case of electrical conductivity, thermal conductivity, semiconducting properties or mechanical strength, needs to be explored in more detail. As presented in this book, the results of experiments and demonstrations of practical applications are promising an emerging application of the additive approach for the mass production of structural electronics. This requires the synergy and knowledge transfer from laboratories and universities to larger production-scale facilities over time to exploit these promising properties fully. There is a great deal of scientific research into the electrical, optical, mechanical or chemical properties of functional materials and development procedures to adapt fabrication techniques for efficient exploitation of these remarkable properties to large-scale production.

The areas for exploration of 3D printed structural electronics are vast. We could dive deeper into the areas not mentioned here, such as detailed investigations on nanomaterials applications in 3D printing, substrate materials modulations and surface activation, micro and nanostructuring, hybrid approaches such as selective electrochemical metallization allowing to achieve metallic conductivity for 3D printed low resistivity composites, resin-based UV cured materials for SLA or direct Inkjet techniques to fabricate conductive structures, precise fabrication of high-frequency radio circuits, broader medical, military or rescue applications, spacecrafts and planet exploration, and many more. Everything mentioned in this book shows examples of breakthrough developments and small successes in the 3D printing of electronics are intended to be a catalyst for out-of-the-box thinking, allowing us to reimagine and redesign our approach to electronics fabrication. If we imagine that we are on the brink of new discoveries and new technology implementation, such as artificial intelligence for materials elaboration and generative designing, with a growing level of automatization and robotization of our industry, and slowly but successively walking into the next industry revolution of being environmentally neutral and energy efficient society, the additive manufacturing of electronics and most of surrounding us goods, making them equipped with more and more smart solutions, will eventually become an every day, ordinary approach for manufacturing, and become a new normal, practical solution for the growth of the society.

Bibliography

[1]   Gordon E. Moore. Cramming more components onto integrated circuits. *Proc. IEEE*, 86(1):82–85, 1998.

[2]   Scott E. Thompson and Srivatsan Parthasarathy. Moore's law: the future of Si microelectronics. *Mater. Today*, 9(6):20–25, June 2006.

[3]   Raymond H. Clark. *Handbook of printed circuit manufacturing*. Springer Netherlands, Dordrecht, 1985.

[4]   Yangyuan Wang, Min-Hwa Chi, Jesse Jen-Chung Lou, and Chun-Zhang Chen. *Handbook of integrated circuit industry*. Springer Nature Singapore, Singapore, 2024.

[5]   William E. Frazier. Metal additive manufacturing: a review. *J. Mater. Eng. Perform.*, 23(6):1917–1928, June 2014.

[6]   Amit Bandyopadhyay and Susmita Bose. *Additive manufacturing*. CRC Press, Boca Raton, 2nd edition, 2020.

[7]   Charles W. Hull. Apparatus for production of three-dimensional objects by stereolithography. US4575330A.

[8]   Scott Crump. Rapid prototyping using FDM. *Mod. Cast.*, 82(4):36–38, 1992.

[9]   Carl R. Deckard. *Part generation by layerwise selective sintering*. MS Dissertation. University of Texas at Austin, 1986.

[10]   Hideo Kodama. Automatic method for fabricating a three-dimensional plastic model with photo-hardening polymer. *Rev. Sci. Instrum.*, 52(11):1770–1773, Nov 1981.

[11]   Richard Ellam. 3D printing: you read it here first. *New Sci.*, (3099), Nov 2016.

[12]   Johannes Gottwald. Liquid metal recorder. https://patents.google.com/patent/US3596285A/en, US3596285A, July 1971.

[13]   Stanisław Lem. *Obłok Magellana*. Iskry, Warsaw, 1955.

[14]   Murray Leinster. *Things pass by*, 1945.

[15]   Thierry Rayna and Ludmila Striukova. From rapid prototyping to home fabrication: how 3D printing is changing business model innovation. *Technol. Forecast. Soc. Change*, 102:214–224, Jan 2016.

[16]   Brian H. Cumpston, Sundaravel P. Ananthavel, Stephen Barlow, Daniel L. Dyer, Jeffrey E. Ehrlich, Lael L. Erskine, Ahmed A. Heikal, Stephen M. Kuebler, I.-Y. Sandy Lee, Dianne McCord-Maughon, Jinqui Qin, Harald Röckel, Mariacristina Rumi, Xiang-Li Wu, Seth R. Marder, and Joseph W. Perry. Two-photon polymerization initiators for three-dimensional optical data storage and microfabrication. *Nature*, 398(6722):51–54, Mar 1999.

[17]   A. Meurisse, A. Makaya, C. Willsch, and M. Sperl. Solar 3D printing of lunar regolith. *Acta Astronaut.*, 152:800–810, Nov 2018.

[18]   Ryan L. Truby and Jennifer A. Lewis. Printing soft matter in three dimensions. *Nature*, 540(7633):371–378, Dec 2016.

[19]   Cheng Zhang, Di Ouyang, Simon Pauly, and Lin Liu. 3D printing of bulk metallic glasses. *Mater. Sci. Eng., R Rep.*, 145:100625, July 2021.

[20]   Byron Blakey-Milner, Paul Gradl, Glen Snedden, Michael Brooks, Jean Pitot, Elena Lopez, Martin Leary, Filippo Berto, and Anton du Plessis. Metal additive manufacturing in aerospace: a review. *Mater. Des.*, 209:110008, Nov 2021.

[21]   Daniel Böckin and Anne-Marie Tillman. Environmental assessment of additive manufacturing in the automotive industry. *J. Clean. Prod.*, 226:977–987, July 2019.

[22]   R. Leal, F. M. Barreiros, L. Alves, F. Romeiro, J. C. Vasco, M. Santos, and C. Marto. Additive manufacturing tooling for the automotive industry. *Int. J. Adv. Manuf. Technol.*, 92(5–8):1671–1676, Sept 2017.

[23]   Joel C. Najmon, Sajjad Raeisi, and Andres Tovar. *Review of additive manufacturing technologies and applications in the aerospace industry*, pp. 7–31. Elsevier, 2019.

[24] Smith Salifu, Dawood Desai, Olugbenga Ogunbiyi, and Kampamba Mwale. Recent development in the additive manufacturing of polymer-based composites for automotive structures—a review. *Int. J. Adv. Manuf. Technol.*, 119(11–12):6877–6891, Apr 2022.

[25] A. A. Shapiro, J. P. Borgonia, Q. N. Chen, R. P. Dillon, B. McEnerney, R. Polit-Casillas, and L. Soloway. Additive manufacturing for aerospace flight applications. *J. Spacecr. Rockets*, 53(5):952–959, Sept 2016.

[26] Nicolas Martelli, Carole Serrano, Hélène van-den Brink, Judith Pineau, Patrice Prognon, Isabelle Borget, and Salma El-Batti. Advantages and disadvantages of 3-dimensional printing in surgery: a systematic review. *Surgery*, 159(6):1485–1500, June 2016.

[27] Sébastien Ruiters, Yi Sun, Stéphan de Jong, Constantinus Politis, and Ilse Mombaerts. Computer-aided design and three-dimensional printing in the manufacturing of an ocular prosthesis. *Br. J. Ophthalmol.*, 100(7):879–881, July 2016.

[28] Jorge Zuniga, Dimitrios Katsavelis, Jean Peck, John Stollberg, Marc Petrykowski, Adam Carson, and Cristina Fernandez. Cyborg beast: a low-cost 3D-printed prosthetic hand for children with upper-limb differences. *BMC Res. Notes*, 8(1):10, 2015.

[29] Alvaro Goyanes, Jie Wang, Asma Buanz, Ramón Martínez-Pacheco, Richard Telford, Simon Gaisford, and Abdul W. Basit. 3D printing of medicines: engineering novel oral devices with unique design and drug release characteristics. *Mol. Pharm.*, 12(11):4077–4084, Nov 2015.

[30] Yong Lin Kong, Xingyu Zou, Caitlin A. McCandler, Ameya R. Kirtane, Shen Ning, Jianlin Zhou, Abubakar Abid, Mousa Jafari, Jaimie Rogner, Daniel Minahan, Joy E. Collins, Shane McDonnell, Cody Cleveland, Taylor Bensel, Siid Tamang, Graham Arrick, Alla Gimbel, Tiffany Hua, Udayan Ghosh, Vance Soares, Nancy Wang, Aniket Wahane, Alison Hayward, Shiyi Zhang, Brian R. Smith, Robert Langer, and Giovanni Traverso. 3D-printed gastric resident electronics. *Adv. Mater. Technol.*, 4(3):1800490, Mar 2019.

[31] Michael A. Luzuriaga, Danielle R. Berry, John C. Reagan, Ronald A. Smaldone, and Jeremiah J. Gassensmith. Biodegradable 3D printed polymer microneedles for transdermal drug delivery. *Lab Chip*, 18(8):1223–1230, 2018.

[32] Iulia D. Ursan, Ligia Chiu, and Andrea Pierce. Three-dimensional drug printing: a structured review. *J. Am. Pharm. Assoc.*, 53(2):136–144, Mar 2013.

[33] Carl Schubert, Mark C van Langeveld, and Larry A Donoso. Innovations in 3D printing: a 3D overview from optics to organs. *Br. J. Ophthalmol.*, 98(2):159–161, Feb 2014.

[34] Satyajit Patra and Vanesa Young. A review of 3D printing techniques and the future in biofabrication of bioprinted tissue. *Cell Biochem. Biophys.*, 74(2):93–98, June 2016.

[35] Youwen Yang, Guoyong Wang, Huixin Liang, Chengde Gao, Shuping Peng, Lida Shen, and Cijun Shuai. Additive manufacturing of bone scaffolds. *Int. J. Bioprinting*, 5(1):148, Dec 2018.

[36] Rod R. Jose, Maria J. Rodriguez, Thomas A. Dixon, Fiorenzo Omenetto, and David L. Kaplan. Evolution of bioinks and additive manufacturing technologies for 3D bioprinting. *ACS Biomater. Sci. Eng.*, 2(10):1662–1678, Oct 2016.

[37] Jin Woo Lee, Yeong-Jin Choi, Woon-Jae Yong, Falguni Pati, Jin-Hyung Shim, Kyung Shin Kang, In-Hye Kang, Jaesung Park, and Dong-Woo Cho. Development of a 3D cell printed construct considering angiogenesis for liver tissue engineering. *Biofabrication*, 8(1):015007, Jan 2016.

[38] Ferry P. W. Melchels, Marco A. N. Domingos, Travis J. Klein, Jos Malda, Paulo J. Bartolo, and Dietmar W. Hutmacher. Additive manufacturing of tissues and organs. *Prog. Polym. Sci.*, 37(8):1079–1104, Aug 2012.

[39] Abhijit Saha, Trevor G. Johnston, Ryan T. Shafranek, Cassandra J. Goodman, Jesse G. Zalatan, Duane W. Storti, Mark A. Ganter, and Alshakim Nelson. Additive manufacturing of catalytically active living materials. *ACS Appl. Mater. Interfaces*, 10(16):13373–13380, Apr 2018.

[40] Jian-Yuan Lee, Jia An, and Chee Kai Chua. Fundamentals and applications of 3D printing for novel materials. *Appl. Mater. Today*, 7:120–133, June 2017.

[41]   C. Y. Yap, C. K. Chua, Z. L. Dong, Z. H. Liu, D. Q. Zhang, L. E. Loh, and S. L. Sing. Review of selective laser melting: materials and applications. *Appl. Phys. Rev.*, 2(4):041101, Dec 2015.

[42]   W. H. Yu, S. L. Sing, C. K. Chua, C. N. Kuo, and X. L. Tian. Particle-reinforced metal matrix nanocomposites fabricated by selective laser melting: a state of the art review. *Prog. Mater. Sci.*, 104:330–379, July 2019.

[43]   Shangqin Yuan, Fei Shen, Chee Kai Chua, and Kun Zhou. Polymeric composites for powder-based additive manufacturing: materials and applications. *Prog. Polym. Sci.*, 91:141–168, Apr 2019.

[44]   Wohlers report 2018: 3D printing and additive manufacturing state of the industry annual worldwide progress report. Wohlers Associates, Fort Collins (Colo.), 2018. ISBN 978-0-9913332-4-0.

[45]   Nikitas Nikitakos, Ioannis Dagkinis, Dimitrios Papachristos, Georgios Georgantis, and Evanthia Kostidi. *Economics in 3D printing*, pp. 85–95. Elsevier, 2020.

[46]   International Organization for Standardization. ISO/ASTM 52900:2021. Additive manufacturing-general principles-fundamentals and vocabulary. https://www.iso.org/obp/ui/#iso:std:iso-astm:52900:ed-2:v1:en.

[47]   P. Dudek. FDM 3D printing technology in manufacturing composite elements. *Arch. Metall. Mater.*, 58(4):1415–1418, Dec 2013.

[48]   David Espalin, Jorge Alberto-Ramirez, Francisco Medina, and Ryan Wicker. Multi-material, multi-technology FDM: exploring build process variations. *Rapid Prototyping J.*, 20(3):236–244, Apr 2014.

[49]   R Melnikova, A Ehrmann, and K Finsterbusch. 3D printing of textile-based structures by fused deposition modelling (FDM) with different polymer materials. *IOP Conf. Ser., Mater. Sci. Eng.*, 62:012018, Aug 2014.

[50]   Bettina Wendel, Dominik Rietzel, Florian Kühnlein, Robert Feulner, Gerrit Hülder, and Ernst Schmachtenberg. Additive processing of polymers: additive processing of polymers. *Macromol. Mater. Eng.*, 293(10):799–809, Oct 2008.

[51]   Freddie Hong, Borut Lampret, Connor Myant, Steve Hodges, and David Boyle. 5-axis multi-material 3D printing of curved electrical traces. *Addit. Manuf.*, 70:103546, May 2023.

[52]   Michael Wüthrich, Wilfried J. Elspass, Philip Bos, and Simon Holdener. *Novel 4-axis 3D printing process to print overhangs without support material*, pp. 130–145. Springer International Publishing, Cham, 2021.

[53]   M. Saravana Kumar, Muhammad Umar Farooq, Nimel Sworna Ross, Che-Hua Yang, V. Kavimani, and Adeolu A. Adediran. Achieving effective interlayer bonding of PLA parts during the material extrusion process with enhanced mechanical properties. *Sci. Rep.*, 13(1):6800, Apr 2023.

[54]   Ruben Bayu Kristiawan, Fitrian Imaduddin, Dody Ariawan, Ubaidillah, and Zainal Arifin. A review on the fused deposition modeling (FDM) 3D printing: filament processing, materials, and printing parameters. *Open Eng.*, 11(1):639–649, Apr 2021.

[55]   Sachini Wickramasinghe, Truong Do, and Phuong Tran. FDM-based 3D printing of polymer and associated composite: a review on mechanical properties, defects and treatments. *Polymers*, 12(7):1529, July 2020.

[56]   Lu-Yu Zhou, Jianzhong Fu, and Yong He. A review of 3D printing technologies for soft polymer materials. *Adv. Funct. Mater.*, 30(28):2000187, 2020.

[57]   Antonio Armillotta. Assessment of surface quality on textured FDM prototypes. *Rapid Prototyping J.*, 12(1):35–41, Jan 2006.

[58]   Reverson Fernandes Quero, Géssica Domingos-da Silveira, José Alberto Fracassi-da Silva, and Dosil Pereira de Jesus. Understanding and improving FDM 3D printing to fabricate high-resolution and optically transparent microfluidic devices. *Lab Chip*, 21(19):3715–3729, 2021.

[59]   I. John Solomon, P. Sevvel, and J. Gunasekaran. A review on the various processing parameters in FDM. *Mater. Today Proc.*, 37:509–514, 2021.

[60]   L. Novakova-Marcincinova, J. Novak-Marcincin, J. Barna, and J. Torok. Special materials used in FDM rapid prototyping technology application. In *2012 IEEE 16th international conference on intelligent engineering systems (INES)*, pp. 73–76, Lisbon, Portugal, June 2012. IEEE.

[61] Antonia Georgopoulou and Frank Clemens. Pellet-based fused deposition modeling for the development of soft compliant robotic grippers with integrated sensing elements. *Flex. Print. Electron.*, 7(2):025010, June 2022.

[62] Narendra Kumar, Prashant Kumar Jain, Puneet Tandon, and Pulak M. Pandey. Additive manufacturing of flexible electrically conductive polymer composites via CNC-assisted fused layer modeling process. *J. Braz. Soc. Mech. Sci. Eng.*, 40(4):175, Apr 2018.

[63] Brian K. Post, Bradley Richardson, Randall Lind, Lonnie J. Love, Peter Lloyd, Vlastimil Kune, Breanna J. Rhyne, Alex Roschli, Jim Hannan, Steve Nolet, Kevin Veloso, Parthiv Kurup, Timothy Remo, and Dale Jenne. Big area additive manufacturing application in wind turbine molds. In *Proceedings of the 20th annual solid freeform fabrication symposium*, 2017.

[64] David Roberson, Corey M Shemelya, Eric MacDonald, and Ryan Wicker. Expanding the applicability of FDM-type technologies through materials development. *Rapid Prototyping J.*, 21(2):137–143, Mar 2015.

[65] Jae Sung Park, Taeil Kim, and Woo Soo Kim. Conductive cellulose composites with low percolation threshold for 3D printed electronics. *Sci. Rep.*, 7(1):3246, June 2017.

[66] B. Podsiadły, A. Skalski, B. Wałpuski, and M. Słoma. Heterophase materials for fused filament fabrication of structural electronics. *J. Mater. Sci., Mater. Electron.*, 30(2):1236–1245, Nov 2018.

[67] Michael Chung, Norbert Radacsi, Colin Robert, Edward D. McCarthy, Anthony Callanan, Noel Conlisk, Peter R. Hoskins, and Vasileios Koutsos. On the optimization of low-cost FDM 3D printers for accurate replication of patient-specific abdominal aortic aneurysm geometry. *3D Print. Med.*, 4(1):2, Dec 2018.

[68] David Baca and Rafiq Ahmad. The impact on the mechanical properties of multi-material polymers fabricated with a single mixing nozzle and multi-nozzle systems via fused deposition modeling. *Int. J. Adv. Manuf. Technol.*, 106(9–10):4509–4520, Feb 2020.

[69] J. A. Lewis. Direct ink writing of 3D functional materials. *Adv. Funct. Mater.*, 16(17):2193–2204, Nov 2006.

[70] Longyu Li, Qianming Lin, Miao Tang, Andrew J. E. Duncan, and Chenfeng Ke. Advanced polymer designs for direct-ink-write 3D printing. *Chem.—Eur. J.*, 25(46):10768–10781, Aug 2019.

[71] M. A. S. R. Saadi, Alianna Maguire, Neethu T. Pottackal, Md Shajedul Hoque Thakur, Maruf Md. Ikram, A. John Hart, Pulickel M. Ajayan, and Muhammad M. Rahman. Direct ink writing: a 3D printing technology for diverse materials. *Adv. Mater.*, 34(28):2108855, July 2022.

[72] Ian D. Robertson, Mostafa Yourdkhani, Polette J. Centellas, Jia En Aw, Douglas G. Ivanoff, Elyas Goli, Evan M. Lloyd, Leon M. Dean, Nancy R. Sottos, Philippe H. Geubelle, Jeffrey S. Moore, and Scott R. White. Rapid energy-efficient manufacturing of polymers and composites via frontal polymerization. *Nature*, 557(7704):223–227, May 2018.

[73] Taylor V. Neumann and Michael D. Dickey. Liquid metal direct write and 3D printing: a review. *Adv. Mater. Technol.*, 5(9):2000070, Sept 2020.

[74] Kaijuan Chen, Lei Zhang, Xiao Kuang, Vincent Li, Ming Lei, Guozheng Kang, Zhong Lin Wang, and Hang Jerry Qi. Dynamic photomask-assisted direct ink writing multimaterial for multilevel triboelectric nanogenerator. *Adv. Funct. Mater.*, 29(33):1903568, Aug 2019.

[75] Young Yoon, Shinmyoung Kim, Doyoon Kim, Sang Ken Kauh, and Jungchul Lee. Four degrees-of-freedom direct writing of liquid metal patterns on uneven surfaces. *Adv. Mater. Technol.*, 4(2):1800379, Feb 2019.

[76] Young-Geun Park, Hyeon Seok An, Ju-Young Kim, and Jang-Ung Park. High-resolution, reconfigurable printing of liquid metals with three-dimensional structures. *Sci. Adv.*, 5(6):eaaw2844, June 2019.

[77] Peikai Zhang, Nihan Aydemir, Maan Alkaisi, David E. Williams, and Jadranka Travas-Sejdic. Direct writing and characterization of three-dimensional conducting polymer PEDOT arrays. *ACS Appl. Mater. Interfaces*, 10(14):11888–11895, Apr 2018.

[78] Bok Yeop Ahn, Steven B. Walker, Scott C. Slimmer, Analisa Russo, Ashley Gupta, Steve Kranz, Eric B. Duoss, Thomas F. Malkowski, and Jennifer A. Lewis. Planar and three-dimensional printing of conductive inks. *J. Vis. Exp.*, (58):3189, Dec 2011.

[79]   Giovanni Pierin, Chiara Grotta, Paolo Colombo, and Cecilia Mattevi. Direct ink writing of micrometric SiOC ceramic structures using a preceramic polymer. *J. Eur. Ceram. Soc.*, 36(7):1589–1594, June 2016.

[80]   Zhenzhong Hou, Hai Lu, Ying Li, Laixia Yang, and Yang Gao. Direct ink writing of materials for electronics-related applications: a mini review. *Front. Mater.*, 8:647229, Apr 2021.

[81]   F. Tricot, C. Venet, D. Beneventi, D. Curtil, D. Chaussy, T. P. Vuong, J. E. Broquin, and N. Reverdy-Bruas. Fabrication of 3D conductive circuits: print quality evaluation of a direct ink writing process. *RSC Adv.*, 8(46):26036–26046, 2018.

[82]   Giorgia Franchin, Larissa Wahl, and Paolo Colombo. Direct ink writing of ceramic matrix composite structures. *J. Am. Ceram. Soc.*, 100(10):4397–4401, Oct 2017.

[83]   T. J. Wallin, J. Pikul, and R. F. Shepherd. 3D printing of soft robotic systems. *Nat. Rev. Mater.*, 3(6):84–100, June 2018.

[84]   Yubai Zhang, Ge Shi, Jiadong Qin, Sean E. Lowe, Shanqing Zhang, Huijun Zhao, and Yu Lin Zhong. Recent progress of direct ink writing of electronic components for advanced wearable devices. *ACS Appl. Electron. Mater.*, 1(9):1718–1734, Sept 2019.

[85]   Guohua Hu, Joohoon Kang, Leonard W. T. Ng, Xiaoxi Zhu, Richard C. T. Howe, Christopher G. Jones, Mark C. Hersam, and Tawfique Hasan. Functional inks and printing of two-dimensional materials. *Chem. Soc. Rev.*, 47(9):3265–3300, 2018.

[86]   Jennifer A Lewis. Direct-write assembly of ceramics from colloidal inks. *Curr. Opin. Solid State Mater. Sci.*, 6(3):245–250, Jun 2002.

[87]   Kathleen Hajash, Bjorn Sparrman, Christophe Guberan, Jared Laucks, and Skylar Tibbits. Large-scale rapid liquid printing. *3D Print. Addit. Manuf.*, 4(3):123–132, Sept 2017.

[88]   Abigail K. Grosskopf, Ryan L. Truby, Hyoungsoo Kim, Antonio Perazzo, Jennifer A. Lewis, and Howard A. Stone. Viscoplastic matrix materials for embedded 3D printing. *ACS Appl. Mater. Interfaces*, 10(27):23353–23361, July 2018.

[89]   Leanne Friedrich and Matthew Begley. In situ characterization of low-viscosity direct ink writing: stability, wetting, and rotational flows. *J. Colloid Interface Sci.*, 529:599–609, Nov 2018.

[90]   Amin M'Barki, Lydéric Bocquet, and Adam Stevenson. Linking rheology and printability for dense and strong ceramics by direct ink writing. *Sci. Rep.*, 7(1):6017, Dec 2017.

[91]   Jordan R. Raney, Brett G. Compton, Jochen Mueller, Thomas J. Ober, Kristina Shea, and Jennifer A. Lewis. Rotational 3D printing of damage-tolerant composites with programmable mechanics. *Proc. Natl. Acad. Sci.*, 115(6):1198–1203, Feb 2018.

[92]   Adam W. Feinberg and Jordan S. Miller. Progress in three-dimensional bioprinting. *Mater. Res. Soc. Bull.*, 42(08):557–562, Aug 2017.

[93]   Marcel Alexander Heinrich, Wanjun Liu, Andrea Jimenez, Jingzhou Yang, Ali Akpek, Xiao Liu, Qingmeng Pi, Xuan Mu, Ning Hu, Raymond Michel Schiffelers, Jai Prakash, Jingwei Xie, and Yu Shrike Zhang. Bioprinting: 3D bioprinting: from benches to translational applications (small 23/2019). *Small*, 15(23):1970126, June 2019.

[94]   Sean V Murphy and Anthony Atala. 3D bioprinting of tissues and organs. *Nat. Biotechnol.*, 32(8):773–785, Aug 2014.

[95]   Joseph T. Muth, Patrick G. Dixon, Logan Woish, Lorna J. Gibson, and Jennifer A. Lewis. Architected cellular ceramics with tailored stiffness via direct foam writing. *Proc. Natl. Acad. Sci.*, 114(8):1832–1837, Feb 2017.

[96]   Manuel Schaffner, Patrick A. Rühs, Fergal Coulter, Samuel Kilcher, and André R. Studart. 3D printing of bacteria into functional complex materials. *Sci. Adv.*, 3(12):eaao6804, Dec 2017.

[97]   Jung Hyun Kim, Won Suk Chang, Daeho Kim, Jong Ryul Yang, Joong Tark Han, Geon-Woong Lee, Ji Tae Kim, and Seung Kwon Seol. 3D printing of reduced graphene oxide nanowires. *Adv. Mater.*, 27(1):157–161, Jan 2015.

[98]   Qiangqiang Zhang, Feng Zhang, Sai Pradeep Medarametla, Hui Li, Chi Zhou, and Dong Lin. 3D printing of graphene aerogels. *Small*, 12(13):1702–1708, Apr 2016.

[99]   Adam E. Jakus, Ethan B. Secor, Alexandra L. Rutz, Sumanas W. Jordan, Mark C. Hersam, and Ramille N. Shah. Three-dimensional printing of high-content graphene scaffolds for electronic and biomedical applications. *ACS Nano*, 9(4):4636–4648, Apr 2015.

[100] Victoria G. Rocha, Eduardo Saiz, Iuliia S. Tirichenko, and Esther García-Tuñón. Direct ink writing advances in multi-material structures for a sustainable future. *J. Mater. Chem. A*, 8(31):15646–15657, 2020.

[101] Osman Dogan Yirmibesoglu, Leif Erik Simonsen, Robert Manson, Joseph Davidson, Katherine Healy, Yigit Menguc, and Thomas Wallin. Multi-material direct ink writing of photocurable elastomeric foams. *Commun. Mater.*, 2(1):82, Dec 2021.

[102] Sean Smyth. *The future of digital printing to 2032*. Smithers Information Ltd, Mar 2022.

[103] F. J. Kamphoefner. Ink jet printing. *IEEE Trans. Electron Devices*, 19(4):584–593, Apr 1972.

[104] G. D. Martin, S. D. Hoath, and I. M. Hutchings. Inkjet printing—the physics of manipulating liquid jets and drops. *J. Phys. Conf. Ser.*, 105:012001, Mar 2008.

[105] Jürgen Brünahl and Alex M. Grishin. Piezoelectric shear mode drop-on-demand inkjet actuator. *Sens. Actuators Phys.*, 101(3):371–382, Oct 2002.

[106] Yang Guo, Huseini S. Patanwala, Brice Bognet, and Anson W. K. Ma. Inkjet and inkjet-based 3D printing: connecting fluid properties and printing performance. *Rapid Prototyping J.*, 23(3):562–576, Apr 2017.

[107] T. C. Gomes, C. J. L. Constantino, E. M. Lopes, A. E. Job, and N. Alves. Thermal inkjet printing of polyaniline on paper. *Thin Solid Films*, 520(24):7200–7204, Oct 2012.

[108] Jerry Fuh. *Micro- and bio-rapid prototyping using drop-on-demand 3D printing*, pp. 1–15. Springer London, London, 2013.

[109] Ian M. Hutchings and Graham D. Martin. *Inkjet technology for digital fabrication*. Wiley, 1st edition, Nov 2012.

[110] Alejandro H. Espera, John Ryan C. Dizon, Qiyi Chen, and Rigoberto C. Advincula. 3D-printing and advanced manufacturing for electronics. *Prog. Addit. Manuf.*, 4(3):245–267, Sept 2019.

[111] Junfeng Mei, M. R. Lovell, and M. H. Mickle. Formulation and processing of novel conductive solution inks in continuous inkjet printing of 3-d electric circuits. *IEEE Trans. Electron. Packag. Manuf.*, 28(3):265–273, July 2005.

[112] Gerard Cummins and Marc P. Y. Desmulliez. Inkjet printing of conductive materials: a review. *Circuit World*, 38(4):193–213, Nov 2012.

[113] Kye-Si Kwon, Md. Khalilur Rahman, Thanh Huy Phung, Steve Hoath, SunHo Jeong, and Jang Sub Kim. Review of digital printing technologies for electronic materials. *Flex. Print. Electron.*, Nov 2020.

[114] Emine Tekin, Patrick J. Smith, and Ulrich S. Schubert. Inkjet printing as a deposition and patterning tool for polymers and inorganic particles. *Soft Matter*, 4(4):703, 2008.

[115] Daehwan Jang, Dongjo Kim, and Jooho Moon. Influence of fluid physical properties on ink-jet printability. *Langmuir*, 25(5):2629–2635, Mar 2009.

[116] Yun Lu Tee, Chenxi Peng, Philip Pille, Martin Leary, and Phuong Tran. Polyjet 3D printing of composite materials: experimental and modelling approach. *JOM*, 72(3):1105–1117, Mar 2020.

[117] Hongyi Yang, Jingying Charlotte Lim, Yuchan Liu, Xiaoying Qi, Yee Ling Yap, Vishwesh Dikshit, Wai Yee Yeong, and Jun Wei. Performance evaluation of projet multi-material jetting 3D printer. *Virtual Phys. Prototyp.*, 12(1):95–103, Jan 2017.

[118] Sarah Krainer, Chris Smit, and Ulrich Hirn. The effect of viscosity and surface tension on inkjet printed picoliter dots. *RSC Adv.*, 9(54):31708–31719, 2019.

[119] Parth Patpatiya, Kailash Chaudhary, Anshuman Shastri, and Shailly Sharma. A review on polyjet 3D printing of polymers and multi-material structures. *Proc. Inst. Mech. Eng., Part C, J. Mech. Eng. Sci.*, 236(14):7899–7926, July 2022.

[120] Ayoung Lee, Kai Sudau, Kyung Hyun Ahn, Seung Jong Lee, and Norbert Willenbacher. Optimization of experimental parameters to suppress nozzle clogging in inkjet printing. *Ind. Eng. Chem. Res.*, 51(40):13195–13204, Oct 2012.

[121] H. W. Tan, T. Tran, and C. K. Chua. A review of printed passive electronic components through fully additive manufacturing methods. *Virtual Phys. Prototyp.*, 11(4):271–288, Oct 2016.

[122] Matic Krivec, Martin Lenzhofer, Thomas Moldaschl, Jaka Pribošek, Anže Abram, and Michael Ortner. Inkjet printing of multi-layered, via-free conductive coils for inductive sensing applications. *Microsyst. Technol.*, 24(6):2673–2682, June 2018.

[123] L. Setti, A. Fraleoni-Morgera, B. Ballarin, A. Filippini, D. Frascaro, and C. Piana. An amperometric glucose biosensor prototype fabricated by thermal inkjet printing. *Biosens. Bioelectron.*, 20(10):2019–2026, Apr 2005.

[124] T. H. J. van Osch, J. Perelaer, A. W. M. de Laat, and U. S. Schubert. Inkjet printing of narrow conductive tracks on untreated polymeric substrates. *Adv. Mater.*, 20(2):343–345, Jan 2008.

[125] Byung-Hun Kim, Hwa-Sun Lee, Sung-Wook Kim, Piljoong Kang, and Yoon-Sok Park. Hydrodynamic responses of a piezoelectric driven MEMS inkjet print-head. *Sens. Actuators Phys.*, 210:131–140, Apr 2014.

[126] Bruce H. King, Michael J. O'Reilly, and Stephen M. Barnes. Characterizing aerosol jet multi-nozzle process parameters for non-contact front side metallization of silicon solar cells. In *2009 34th IEEE photovoltaic specialists conference (PVSC)*, pp. 001107–001111, Philadelphia, PA, USA, Jun 2009. IEEE.

[127] Allon Shimoni, Suzanna Azoubel, and Shlomo Magdassi. Inkjet printing of flexible high-performance carbon nanotube transparent conductive films by "coffee ring effect". *Nanoscale*, 6(19):11084–11089, 2014.

[128] Dan Soltman and Vivek Subramanian. Inkjet-printed line morphologies and temperature control of the coffee ring effect. *Langmuir*, 24(5):2224–2231, Mar 2008.

[129] Justin M. Hoey, Artur Lutfurakhmanov, Douglas L. Schulz, and Iskander S. Akhatov. A review on aerosol-based direct-write and its applications for microelectronics. *J. Nanotechnol.*, 2012:1–22, 2012.

[130] N. J. Wilkinson, M. A. A. Smith, R. W. Kay, and R. A. Harris. A review of aerosol jet printing—a non-traditional hybrid process for micro-manufacturing. *Int. J. Adv. Manuf. Technol.*, 105(11):4599–4619, Dec 2019.

[131] Ankit Mahajan, C. Daniel Frisbie, and Lorraine F. Francis. Optimization of aerosol jet printing for high-resolution, high-aspect ratio silver lines. *ACS Appl. Mater. Interfaces*, 5(11):4856–4864, June 2013.

[132] Ethan B. Secor. Principles of aerosol jet printing. *Flex. Print. Electron.*, 3(3):035002, Sep 2018.

[133] Jeffrey G. Tait, Ewelina Witkowska, Masaya Hirade, Tung-Huei Ke, Pawel E. Malinowski, Soeren Steudel, Chihaya Adachi, and Paul Heremans. Uniform aerosol jet printed polymer lines with 30 μm width for 140ppi resolution rgb organic light emitting diodes. *Org. Electron.*, 22:40–43, July 2015.

[134] Marcel Rother, Maximilian Brohmann, Shuyi Yang, Stefan B. Grimm, Stefan P. Schießl, Arko Graf, and Jana Zaumseil. Aerosol-jet printing of polymer-sorted (6,5) carbon nanotubes for field-effect transistors with high reproducibility. *Adv. Electron. Mater.*, 3(8):1700080, Aug 2017.

[135] Christian E. Folgar, Carlos Suchicital, and Shashank Priya. Solution-based aerosol deposition process for synthesis of multilayer structures. *Mater. Lett.*, 65(9):1302–1307, May 2011.

[136] Ingo Grunwald, Esther Groth, Ingo Wirth, Julian Schumacher, Marcus Maiwald, Volker Zoellmer, and Matthias Busse. Surface biofunctionalization and production of miniaturized sensor structures using aerosol printing technologies. *Biofabrication*, 2(1):014106, Mar 2010.

[137] Jason A. Paulsen, Michael Renn, Kurt Christenson, and Richard Plourde. Printing conformal electronics on 3D structures with aerosol jet technology. In *2012 Future of instrumentation international workshop (FIIW) proceedings*, pp. 1–4, Gatlinburg, TN, USA, Oct 2012. IEEE.

[138] Brian Elder, Rajan Neupane, Eric Tokita, Udayan Ghosh, Samuel Hales, and Yong Lin Kong. Nanomaterial patterning in 3D printing. *Adv. Mater.*, 32(17):1907142, Apr 2020.

[139] Kihyon Hong, Yong Hyun Kim, Se Hyun Kim, Wei Xie, Weichao David Xu, Chris H. Kim, and C. Daniel Frisbie. Aerosol jet printed, sub-2 v complementary circuits constructed from p - and n -type electrolyte gated transistors. *Adv. Mater.*, 26(41):7032–7037, Nov 2014.

[140] AJP application videos. https://www.optomec.com/resources/3d-printing-application-videos/. Accessed: 2022-11-27.

[141] Ali Bagheri and Jianyong Jin. Photopolymerization in 3D printing. *ACS Appl. Polym. Mater.*, 1(4):593–611, Apr 2019.

[142] Erin M. Maines, Mayuri K. Porwal, Christopher J. Ellison, and Theresa M. Reineke. Sustainable advances in SLA/DLP 3D printing materials and processes. *Green Chem.*, 23(18):6863–6897, 2021.

[143] Yayue Pan, Yong Chen, and Chi Zhou. Fast recoating methods for the projection-based stereolithography process in micro- and macro-scales. In *Proceedings of the 23th annual solid freeform fabrication symposium*, 2012.

[144] Sergio Rossi, Alessandra Puglisi, and Maurizio Benaglia. Additive manufacturing technologies: 3D printing in organic synthesis. *ChemCatChem*, 10(7):1512–1525, Apr 2018.

[145] Chenming Wu, Ran Yi, Yong-Jin Liu, Ying He, and Charlie C. L. Wang. Delta DLP 3D printing with large size. In *2016 IEEE/RSJ international conference on intelligent robots and systems (IROS)*, pp. 2155–2160, Daejeon, South Korea, Oct 2016. IEEE.

[146] C. Sun, N. Fang, D. M. Wu, and X. Zhang. Projection micro-stereolithography using digital micro-mirror dynamic mask. *Sens. Actuators Phys.*, 121(1):113–120, May 2005.

[147] Xiaoyu Zheng, Joshua Deotte, Matthew P. Alonso, George R. Farquar, Todd H. Weisgraber, Steven Gemberling, Howon Lee, Nicholas Fang, and Christopher M. Spadaccini. Design and optimization of a light-emitting diode projection micro-stereolithography three-dimensional manufacturing system. *Rev. Sci. Instrum.*, 83(12):125001, Dec 2012.

[148] Rima Janusziewicz, John R. Tumbleston, Adam L. Quintanilla, Sue J. Mecham, and Joseph M. DeSimone. Layerless fabrication with continuous liquid interface production. *Proc. Natl. Acad. Sci.*, 113(42):11703–11708, Oct 2016.

[149] John R. Tumbleston, David Shirvanyants, Nikita Ermoshkin, Rima Janusziewicz, Ashley R. Johnson, David Kelly, Kai Chen, Robert Pinschmidt, Jason P. Rolland, Alexander Ermoshkin, Edward T. Samulski, and Joseph M. DeSimone. Continuous liquid interface production of 3D objects. *Science*, 347(6228):1349–1352, Mar 2015.

[150] Kotaro Obata, Ayman El-Tamer, Lothar Koch, Ulf Hinze, and Boris N. Chichkov. High-aspect 3D two-photon polymerization structuring with widened objective working range (wow-2pp). *Light: Sci. Appl.*, 2(12):e116, Dec 2013.

[151] Shoji Maruo, Osamu Nakamura, and Satoshi Kawata. Three-dimensional microfabrication with two-photon-absorbed photopolymerization. *Opt. Lett.*, 22(2):132, Jan 1997.

[152] V. Ferreras Paz, M. Emons, K. Obata, A. Ovsianikov, S. Peterhänsel, K. Frenner, C. Reinhardt, B. Chichkov, U. Morgner, and W. Osten. Development of functional sub-100 nm structures with 3D two-photon polymerization technique and optical methods for characterization. *J. Laser Appl.*, 24(4):042004, Sept 2012.

[153] Indrasen Bhattacharya, Joseph Toombs, and Hayden Taylor. High fidelity volumetric additive manufacturing. *Addit. Manuf.*, 47:102299, Nov 2021.

[154] Brett E. Kelly, Indrasen Bhattacharya, Hossein Heidari, Maxim Shusteff, Christopher M. Spadaccini, and Hayden K. Taylor. Volumetric additive manufacturing via tomographic reconstruction. *Science*, 363(6431):1075–1079, Mar 2019.

[155] Young-Geun Park, Insik Yun, Won Gi Chung, Wonjung Park, Dong Ha Lee, and Jang-Ung Park. High-resolution 3D printing for electronics. *Adv. Sci.*, 9(8):2104623, Mar 2022.

[156] Boris Metral, Adrien Bischoff, Christian Ley, Ahmad Ibrahim, and Xavier Allonas. Photochemical study of a three-component photocyclic initiating system for free radical photopolymerization: implementing a model for digital light processing 3D printing. *ChemPhotoChem*, 3(11):1109–1118, Nov 2019.

[157] Setareh Zakeri, Minnamari Vippola, and Erkki Levänen. A comprehensive review of the photopolymerization of ceramic resins used in stereolithography. *Addit. Manuf.*, 35:101177, Oct 2020.

[158] Xiaohang Sun, Preeti Tyagi, Sachin Agate, Marian G. McCord, Lucian A. Lucia, and Lokendra Pal. Highly tunable bioadhesion and optics of 3D printable PNIPAm/cellulose nanofibrils hydrogels. *Carbohydr. Polym.*, 234:115898, Apr 2020.

[159] Dong Lin, Shengyu Jin, Feng Zhang, Chao Wang, Yiqian Wang, Chi Zhou, and Gary J Cheng. 3D stereolithography printing of graphene oxide reinforced complex architectures. *Nanotechnology*, 26(43):434003, Oct 2015.

[160] Henry H. Hwang, Wei Zhu, Grace Victorine, Natalie Lawrence, and Shaochen Chen. 3D-printing of functional biomedical microdevices via light- and extrusion-based approaches. *Small Methods*, 2(2):1700277, Feb 2018.

[161] L. E. Murr. Open-cellular metal implant design and fabrication for biomechanical compatibility with bone using electron beam melting. *J. Mech. Behav. Biomed. Mater.*, 76:164–177, Dec 2017.

[162] Amy X. Y. Guo, Liangjie Cheng, Shuai Zhan, Shouyang Zhang, Wei Xiong, Zihan Wang, Gang Wang, and Shan Cecilia Cao. Biomedical applications of the powder-based 3D printed titanium alloys: a review. *J. Mater. Sci. Technol.*, 125:252–264, Oct 2022.

[163] Mattia Mele, Giampaolo Campana, and Gian Luca Monti. Modelling of the capillarity effect in multi jet fusion technology. *Addit. Manuf.*, 30:100879, Dec 2019.

[164] Amir Mostafaei, Amy M. Elliott, John E. Barnes, Fangzhou Li, Wenda Tan, Corson L. Cramer, Peeyush Nandwana, and Markus Chmielus. Binder jet 3D printing—process parameters, materials, properties, modeling, and challenges. *Prog. Mater. Sci.*, 119:100707, June 2021.

[165] Seyed Farid Seyed Shirazi, Samira Gharehkhani, Mehdi Mehrali, Hooman Yarmand, Hendrik Simon Cornelis Metselaar, Nahrizul Adib Kadri, and Noor Azuan Abu Osman. A review on powder-based additive manufacturing for tissue engineering: selective laser sintering and inkjet 3D printing. *Sci. Technol. Adv. Mater.*, 16(3):033502, June 2015.

[166] An Wang, Hongze Wang, Yi Wu, and Haowei Wang. 3D printing of aluminum alloys using laser powder deposition: a review. *Int. J. Adv. Manuf. Technol.*, 116(1–2):1–37, Sept 2021.

[167] Peeyush Nandwana, Amy M. Elliott, Derek Siddel, Abbey Merriman, William H. Peter, and Sudarsanam S. Babu. Powder bed binder jet 3D printing of inconel 718: densification, microstructural evolution and challenges. *Curr. Opin. Solid State Mater. Sci.*, 21(4):207–218, Aug 2017.

[168] Hp metal jet s100 3D printing solution. https://www.hp.com/us-en/printers/3d-printers/products/metal-jet.html. Accessed: 2022-11-27.

[169] Joseph Ingaglio, John Fox, Clay J. Naito, and Paolo Bocchini. Material characteristics of binder jet 3D printed hydrated CSA cement with the addition of fine aggregates. *Constr. Build. Mater.*, 206:494–503, May 2019.

[170] Michael Feygin and Sung Pak. Apparatus for forming an integral object from laminations. https://patents.google.com/patent/US5637175A/en, US5637175A.

[171] Don Klosterman, Richard Chartoff, Nora Osborne, and George Graves. *Laminated object manufacturing, a new process for the direct manufacture of monolithic ceramics and continuous fiber CMCs*, volume 18, pp. 112–120. John Wiley and Sons, Inc., Hoboken, NJ, USA, Jan 1997.

[172] L. Weisensel, N. Travitzky, H. Sieber, and P. Greil. Laminated object manufacturing (LOM) of SISIC composites. *Adv. Eng. Mater.*, 6(11):899–903, Nov 2004.

[173] Manu K. Mohan, A. V. Rahul, Geert De-Schutter, and Kim Van Tittelboom. Extrusion-based concrete 3D printing from a material perspective: a state-of-the-art review. *Cem. Concr. Compos.*, 115:103855, Jan 2021.

[174] Jingchuan Zhang, Jialiang Wang, Sufen Dong, Xun Yu, and Baoguo Han. A review of the current progress and application of 3D printed concrete. *Composites, Part A, Appl. Sci. Manuf.*, 125:105533, Oct 2019.

[175] Clayton Greer, Andrzej Nycz, Mark Noakes, Brad Richardson, Brian Post, Thomas Kurfess, and Lonnie Love. Introduction to the design rules for metal big area additive manufacturing. *Addit. Manuf.*, 27:159–166, May 2019.

[176] Ling Li, Angelica Tirado, I. C. Nlebedim, Orlando Rios, Brian Post, Vlastimil Kunc, R. R. Lowden, Edgar Lara-Curzio, Robert Fredette, John Ormerod, Thomas A. Lograsso, and M. Parans Paranthaman. Big area additive manufacturing of high performance bonded NdFeB magnets. *Sci. Rep.*, 6(1):36212, Dec 2016.

[177] Alex Roschli, Katherine T. Gaul, Alex M. Boulger, Brian K. Post, Phillip C. Chesser, Lonnie J. Love, Fletcher Blue, and Michael Borish. Designing for big area additive manufacturing. *Addit. Manuf.*, 25:275–285, Jan 2019.

[178] Sudhanshu Ranjan Singh and Pradeep Khanna. Wire arc additive manufacturing (WAAM): a new process to shape engineering materials. *Mater. Today Proc.*, 44:118–128, 2021.

[179] Fude Wang, Stewart Williams, Paul Colegrove, and Alphons A. Antonysamy. Microstructure and mechanical properties of wire and arc additive manufactured Ti-6Al-4V. *Metall. Mater. Trans. A*, 44(2):968–977, Feb 2013.

[180] Do-Sik Shim, Gyeong-Yun Baek, Jin-Seon Seo, Gwang-Yong Shin, Kee-Poong Kim, and Ki-Yong Lee. Effect of layer thickness setting on deposition characteristics in direct energy deposition (DED) process. *Opt. Laser Technol.*, 86:69–78, Dec 2016.

[181] Ming Fang, Sanjeev Chandra, and Chul B. Park. Building three-dimensional objects by deposition of molten metal droplets. *Rapid Prototyping J.*, 14(1):44–52, Jan 2008.

[182] Bartłomiej Podsiadły, Liubomir Bezgan, and Marcin Słoma. 3D printed electronic circuits from fusible alloys. *Electronics*, 11(22):3829, Nov 2022.

[183] V. Sukhotskiy, I. H. Karampelas, G. Garg, A. Verma, M. Tong, S. Vader, Z. Vader, and E. P. Furlani. Magnetohydrodynamic drop-on-demand liquid metal 3D printing. In *Proceedings of the 28th annual solid freeform fabrication symposium*, 2017.

[184] Chien-Hsun Wang, Ho-Lin Tsai, Yu-Che Wu, and Weng-Sing Hwang. Investigation of molten metal droplet deposition and solidification for 3D printing techniques. *J. Micromech. Microeng.*, 26(9):095012, Sept 2016.

[185] Imposible Objects. CBAM printer. https://www.impossible-objects.com/cbam-printer/.

[186] Zheng Cui. *Printed electronics: materials, technologies and applications*. Wiley/Higher Education Press, Hoboken, NJ: Solaris South Tower, Singapore, 2016.

[187] Katsuaki Suganuma. *Introduction to printed electronics*. SpringerBriefs in electrical and computer engineering. Springer Science + Business Media, New York, 2014.

[188] Matthew Dyson. *Flexible and printed electronics 2023–2033: forecasts, technologies, markets*. June 2023. https://www.idtechex.com/en/research-report/flexible-and-printed-electronics-2023-2033-forecasts-technologies-markets/943. ISBN 978-1-915514-71-4.

[189] Diana Gregor-Svetec. *Intelligent packaging*, pp. 203–247. Elsevier, 2018.

[190] Xuan Cao, Christian Lau, Yihang Liu, Fanqi Wu, Hui Gui, Qingzhou Liu, Yuqiang Ma, Haochuan Wan, Moh. R. Amer, and Chongwu Zhou. Fully screen-printed, large-area, and flexible active-matrix electrochromic displays using carbon nanotube thin-film transistors. *ACS Nano*, 10(11):9816–9822, Nov 2016.

[191] Frederik C. Krebs, Jan Fyenbo, and Mikkel Jørgensen. Product integration of compact roll-to-roll processed polymer solar cell modules: methods and manufacture using flexographic printing, slot-die coating and rotary screen printing. *J. Mater. Chem.*, 20(41):8994, 2010.

[192] Frederik C. Krebs, Mikkel Jørgensen, Kion Norrman, Ole Hagemann, Jan Alstrup, Torben D. Nielsen, Jan Fyenbo, Kaj Larsen, and Jette Kristensen. A complete process for production of flexible large area polymer solar cells entirely using screen printing—first public demonstration. *Sol. Energy Mater. Sol. Cells*, 93(4):422–441, Apr 2009.

[193] Heera Menon, Remadevi Aiswarya, and Kuzhichalil Peethambharan Surendran. Screen printable MWCNT inks for printed electronics. *RSC Adv.*, 7(70):44076–44081, 2017.

[194] Fanny Hoeng, Julien Bras, Erwan Gicquel, Guillaume Krosnicki, and Aurore Denneulin. Inkjet printing of nanocellulose–silver ink onto nanocellulose coated cardboard. *RSC Adv.*, 7(25):15372–15381, 2017.

[195] Xianzhong Lin, Jaison Kavalakkatt, Martha Ch. Lux-Steiner, and Ahmed Ennaoui. Inkjet-printed Cu 2 ZnSn(S, Se) 4 solar cells. *Adv. Sci.*, 2(6):1500028, June 2015.

[196] Ana Moya, Gemma Gabriel, Rosa Villa, and F. Javier-del Campo. Inkjet-printed electrochemical sensors. *Curr. Opin. Electrochem.*, 3(1):29–39, June 2017.

[197] Krishna Rao R. Venkata, Abhinav K. Venkata, P. S. Karthik, and Surya Prakash Singh. Conductive silver inks and their applications in printed and flexible electronics. *RSC Adv.*, 5(95):77760–77790, 2015.

[198] Nan Zhang, Jing Luo, Ren Liu, and Xiaoya Liu. Tannic acid stabilized silver nanoparticles for inkjet printing of conductive flexible electronics. *RSC Adv.*, 6(87):83720–83729, 2016.

[199] Michal Kerndl and Pavel Steffan. Usage of offset printing technology for printed electronics and smart labels. In *2020 43rd international conference on telecommunications and signal processing (TSP)*, pp. 637–639, Milan, Italy, July 2020. IEEE.

[200] Yasuyuki Kusaka, Nobuko Fukuda, and Hirobumi Ushijima. Recent advances in reverse offset printing: an emerging process for high-resolution printed electronics. *Jpn. J. Appl. Phys.*, 59(SG):SG0802, Apr 2020.

[201] A. C. Hübler, G. C. Schmidt, H. Kempa, K. Reuter, M. Hambsch, and M. Bellmann. Three-dimensional integrated circuit using printed electronics. *Org. Electron.*, 12(3):419–423, Mar 2011.

[202] Z. W. Zhong, J. H. Ee, S. H. Chen, and X. C. Shan. Parametric investigation of flexographic printing processes for R2R printed electronics. *Mater. Manuf. Process.*, 35(5):564–571, Apr 2020.

[203] Xuan Binh Cao, Le Phuong Hoang, Cuc Nguyen Thi Kim, and Toan Thang Vu. Laser ablation on coated metal gravures for roll-to-roll printed electronics. *Opt. Commun.*, 527:128948, Jan 2023.

[204] Terho Kololuoma, Jaakko Leppäniemi, Himadri Majumdar, Rita Branquinho, Elena Herbei-Valcu, Viorica Musat, Rodrigo Martins, Elvira Fortunato, and Ari Alastalo. Gravure printed sol-gel derived alooh hybrid nanocomposite thin films for printed electronics. *J. Mater. Chem. C*, 3(8):1776–1786, 2015.

[205] Hongli Zhu, Binu Baby Narakathu, Zhiqiang Fang, Ahmed Tausif-Aijazi, Margaret Joyce, Massood Atashbar, and Liangbing Hu. A gravure printed antenna on shape-stable transparent nanopaper. *Nanoscale*, 6(15):9110, June 2014.

[206] Vimanyu Beedasy and Patrick J. Smith. Printed electronics as prepared by inkjet printing. *Materials*, 13(3):704, Feb 2020.

[207] Giovanni Nisato, Donald Lupo, Simone Ganz, editors. *Printing and processing techniques*. Jenny Stanford Publishing, 2016.

[208] Ethan B. Secor, Nelson S. Bell, Monica Presiliana Romero, Rebecca R. Tafoya, Thao H. Nguyen, and Timothy J. Boyle. Titanium hydride nanoparticles and nanoinks for aerosol jet printed electronics. *Nanoscale*, 14(35):12651–12657, 2022.

[209] Giovanni Nisato, Donald Lupo, Simone Ganz, editors. *Organic and printed electronics: fundamentals and applications*. Pan Stanford Publishing, Singapore, 2016.

[210] Andrea C. Ferrari, Francesco Bonaccorso, Vladimir Fal'ko, Konstantin S. Novoselov, Stephan Roche, Peter Bøggild, Stefano Borini, Frank H. L. Koppens, Vincenzo Palermo, Nicola Pugno, José A. Garrido, Roman Sordan, Alberto Bianco, Laura Ballerini, Maurizio Prato, Elefterios Lidorikis, Jani Kivioja, Claudio Marinelli, Tapani Ryhänen, Alberto Morpurgo, Jonathan N. Coleman, Valeria Nicolosi, Luigi Colombo, Albert Fert, Mar Garcia-Hernandez, Adrian Bachtold, Grégory F. Schneider, Francisco Guinea, Cees Dekker, Matteo Barbone, Zhipei Sun, Costas Galiotis, Alexander N. Grigorenko, Gerasimos Konstantatos, Andras Kis, Mikhail Katsnelson, Lieven Vandersypen, Annick Loiseau, Vittorio Morandi, Daniel Neumaier, Emanuele Treossi, Vittorio Pellegrini, Marco Polini, Alessandro Tredicucci, Gareth M. Williams, Byung Hee Hong, Jong-Hyun Ahn, Jong Min Kim, Herbert Zirath, Bart J. van Wees, Herre van der Zant, Luigi Occhipinti, Andrea Di Matteo, Ian A. Kinloch, Thomas Seyller, Etienne Quesnel, Xinliang Feng, Ken Teo, Nalin Rupesinghe, Pertti Hakonen, Simon R. T. Neil, Quentin Tannock, Tomas Löfwander, and Jari Kinaret. Science and technology roadmap for graphene, related two-dimensional crystals, and hybrid systems. *Nanoscale*, 7(11):4598–4810, 2015.

[211] Felice Torrisi, Tawfique Hasan, Weiping Wu, Zhipei Sun, Antonio Lombardo, Tero S. Kulmala, Gen-Wen Hsieh, Sungjune Jung, Francesco Bonaccorso, Philip J. Paul, Daping Chu, and Andrea C. Ferrari. Inkjet-printed graphene electronics. *ACS Nano*, 6(4):2992–3006, Apr 2012.

[212] V. Correia, K. Y. Mitra, H. Castro, J. G. Rocha, E. Sowade, R. R. Baumann, and S. Lanceros-Mendez. Design and fabrication of multilayer inkjet-printed passive components for printed electronics circuit development. *J. Manuf. Process.*, 31:364–371, Jan 2018.

[213] Wataru Honda, Shingo Harada, Takayuki Arie, Seiji Akita, and Kuniharu Takei. Wearable, human-interactive, health-monitoring, wireless devices fabricated by macroscale printing techniques. *Adv. Funct. Mater.*, 24(22):3299–3304, June 2014.

[214] Vivek Subramanian, Josephine B. Chang, Alejandro De La Fuente Vornbrock, Daniel C. Huang, Lakshmi Jagannathan, Frank Liao, Brian Mattis, Steven Molesa, David R. Redinger, Daniel Soltman, Steven K. Volkman, and Qintao Zhang. Printed electronics for low-cost electronic systems: technology status and application development. In *ESSCIRC 2008—34th European solid-state circuits conference*, pp. 17–24, Edinburgh, UK, Sept 2008. IEEE.

[215] Robert Abbel, Pit Teunissen, Eric Rubingh, Tim van Lammeren, Romain Cauchois, Marcel Everaars, Joost Valeton, Sjoerd van de Geijn, and Pim Groen. Industrial-scale inkjet printed electronics manufacturing—production up-scaling from concept tools to a roll-to-roll pilot line. *Transl. Mater. Res.*, 1(1):015002, July 2014.

[216] Alexander Kamyshny and Shlomo Magdassi. Conductive nanomaterials for 2D and 3D printed flexible electronics. *Chem. Soc. Rev.*, 48(6):1712–1740, 2019.

[217] Michael Layani, Alexander Kamyshny, and Shlomo Magdassi. Transparent conductors composed of nanomaterials. *Nanoscale*, 6(11):5581–5591, 2014.

[218] Ken Gilleo. *Polymer thick film*. Van Nostrand Reinhold, New York, 1996.

[219] Dong-Hau Kuo, Chien-Chih Chang, Te-Yeu Su, Wun-Ku Wang, and Bin-Yuan Lin. Dielectric behaviours of multi-doped BaTiO3/epoxy composites. *J. Eur. Ceram. Soc.*, 21(9):1171–1177, Sept 2001.

[220] K. Russel. Copper polymer thick film for high density and multilayer interconnects. In *Electron. packag. prod.*, pp. 58–63, 1991.

[221] Alexander Kamyshny and Shlomo Magdassi. Conductive nanomaterials for printed electronics. *Small*, 10(17):3515–3535, Sept 2014.

[222] Sanghyeok Kim, Sejeong Won, Gi-Dong Sim, Inkyu Park, and Soon-Bok Lee. Tensile characteristics of metal nanoparticle films on flexible polymer substrates for printed electronics applications. *Nanotechnology*, 24(8):085701, Mar 2013.

[223] Daniel Langley, Gaël Giusti, Céline Mayousse, Caroline Celle, Daniel Bellet, and Jean-Pierre Simonato. Flexible transparent conductive materials based on silver nanowire networks: a review. *Nanotechnology*, 24(45):452001, Nov 2013.

[224] David McCoul, Weili Hu, Mengmeng Gao, Vishrut Mehta, and Qibing Pei. Recent advances in stretchable and transparent electronic materials. *Adv. Electron. Mater.*, 2(5):1500407, May 2016.

[225] Hye Jin Park, Yejin Jo, Min Kyung Cho, Jeong Young-Woo, Dojin Kim, Su Yeon Lee, Youngmin Choi, and Sunho Jeong. Highly durable Cu-based electrodes from a printable nanoparticle mixture ink: flash-light-sintered, kinetically-controlled microstructure. *Nanoscale*, 10(11):5047–5053, 2018.

[226] B. Reiser, L. González-García, I. Kanelidis, J. H. M. Maurer, and T. Kraus. Gold nanorods with conjugated polymer ligands: sintering-free conductive inks for printed electronics. *Chem. Sci.*, 7(7):4190–4196, 2016.

[227] Yeon-Ho Son, Joon-Young Jang, Min Kyu Kang, Sunghoon Ahn, and Caroline Sunyong Lee. Application of flash-light sintering method to flexible inkjet printing using anti-oxidant copper nanoparticles. *Thin Solid Films*, 656:61–67, June 2018.

[228] Wei Wu. Inorganic nanomaterials for printed electronics: a review. *Nanoscale*, 9(22):7342–7372, 2017.

[229] Le Cai and Chuan Wang. Carbon nanotube flexible and stretchable electronics. *Nanoscale Res. Lett.*, 10(1):320, Dec 2015.

[230] Yuchi Che, Haitian Chen, Hui Gui, Jia Liu, Bilu Liu, and Chongwu Zhou. Review of carbon nanotube nanoelectronics and macroelectronics. *Semicond. Sci. Technol.*, 29(7):073001, July 2014.

[231] K. D. Harris, A. L. Elias, and H.-J. Chung. Flexible electronics under strain: a review of mechanical characterization and durability enhancement strategies. *J. Mater. Sci.*, 51(6):2771–2805, Mar 2016.

[232] Krisztián Kordás, Tero Mustonen, Géza Tóth, Heli Jantunen, Marja Lajunen, Caterina Soldano, Saikat Talapatra, Swastik Kar, Robert Vajtai, and Pulickel M. Ajayan. Inkjet printing of electrically conductive patterns of carbon nanotubes. *Small*, 2(8–9):1021–1025, Aug 2006.

[233] Peter N. Nirmalraj, Philip E. Lyons, Sukanta De, Jonathan N. Coleman, and John J. Boland. Electrical connectivity in single-walled carbon nanotube networks. *Nano Lett.*, 9(11):3890–3895, Nov 2009.

[234] Steve Park, Michael Vosguerichian, and Zhenan Bao. A review of fabrication and applications of carbon nanotube film-based flexible electronics. *Nanoscale*, 5(5):1727, 2013.

[235] Jana Zaumseil. Single-walled carbon nanotube networks for flexible and printed electronics. *Semicond. Sci. Technol.*, 30(7):074001, July 2015.

[236] Wenbo Li, Fengyu Li, Huizeng Li, Meng Su, Meng Gao, Yanan Li, Dan Su, Xingye Zhang, and Yanlin Song. Flexible circuits and soft actuators by printing assembly of graphene. *ACS Appl. Mater. Interfaces*, 8(19):12369–12376, May 2016.

[237] K. S. Novoselov, V. I. Fal'ko, L. Colombo, P. R. Gellert, M. G. Schwab, and K. Kim. A roadmap for graphene. *Nature*, 490(7419):192–200, Oct 2012.

[238] Marc H. Overgaard, Martin Kühnel, Rasmus Hvidsten, Søren V. Petersen, Tom Vosch, Kasper Nørgaard, and Bo W. Laursen. Highly conductive semitransparent graphene circuits screen-printed from water-based graphene oxide ink. *Adv. Mater. Technol.*, 2(7):1700011, July 2017.

[239] Ethan B. Secor, Sooman Lim, Heng Zhang, C. Daniel Frisbie, Lorraine F. Francis, and Mark C. Hersam. Gravure printing of graphene for large-area flexible electronics. *Adv. Mater.*, 26(26):4533–4538, July 2014.

[240] Dong-Ming Sun, Chang Liu, Wen-Cai Ren, and Hui-Ming Cheng. A review of carbon nanotube- and graphene-based flexible thin-film transistors. *Small*, 9(8):1188–1205, Apr 2013.

[241] W. Clemens, W. Fix, J. Ficker, A. Knobloch, and A. Ullmann. From polymer transistors toward printed electronics. *J. Mater. Res.*, 19(7):1963–1973, July 2004.

[242] B. Weng, R. L. Shepherd, K. Crowley, A. J. Killard, and G. G. Wallace. Printing conducting polymers. *Analyst*, 135(11):2779, 2010.

[243] Chi-Yuan Yang, Marc-Antoine Stoeckel, Tero-Petri Ruoko, Han-Yan Wu, Xianjie Liu, Nagesh B. Kolhe, Ziang Wu, Yuttapoom Puttisong, Chiara Musumeci, Matteo Massetti, Hengda Sun, Kai Xu, Deyu Tu, Weimin M. Chen, Han Young Woo, Mats Fahlman, Samson A. Jenekhe, Magnus Berggren, and Simone Fabiano. A high-conductivity n-type polymeric ink for printed electronics. *Nat. Commun.*, 12(1):2354, Dec 2021.

[244] Eric N. Dattoli and Wei Lu. Ito nanowires and nanoparticles for transparent films. *Mater. Res. Soc. Bull.*, 36(10):782–788, Oct 2011.

[245] Y. Kashiwagi, A. Koizumi, Y. Takemura, S. Furuta, M. Yamamoto, M. Saitoh, M. Takahashi, T. Ohno, Y. Fujiwara, K. Murahashi, K. Ohtsuka, and M. Nakamoto. Direct transparent electrode patterning on layered GaN substrate by screen printing of indium tin oxide nanoparticle ink for Eu-doped GaN red light-emitting diode. *Appl. Phys. Lett.*, 105(22):223509, Dec 2014.

[246] Jizhong Song, Sergei A. Kulinich, Jianhai Li, Yanli Liu, and Haibo Zeng. A general one-pot strategy for the synthesis of high-performance transparent-conducting-oxide nanocrystal inks for all-solution-processed devices. *Angew. Chem.*, 127(2):472–476, Jan 2015.

[247] Areum Kim, Yulim Won, Kyoohee Woo, Chul-Hong Kim, and Jooho Moon. Highly transparent low resistance ZnO/Ag nanowire/ZnO composite electrode for thin film solar cells. *ACS Nano*, 7(2):1081–1091, Feb 2013.

[248] Muhammad Muqeet Rehman, Ghayas Uddin Siddiqui, Jahan Zeb Gul, Soo-Wan Kim, Jong Hwan Lim, and Kyung Hyun Choi. Resistive switching in all-printed, flexible and hybrid MoS2-PVA nanocomposite based memristive device fabricated by reverse offset. *Sci. Rep.*, 6(1):36195, Dec 2016.

[249] Jörg J. Schneider, Rudolf C. Hoffmann, Jörg Engstler, Stefan Dilfer, Andreas Klyszcz, Emre Erdem, Peter Jakes, and Rüdiger A. Eichel. Zinc oxide derived from single source precursor chemistry under chimie douce conditions: formation pathway, defect chemistry and possible applications in thin film printing. *J. Mater. Chem.*, 19(10):1449, 2009.

[250] Sooman Lim, Byungjin Cho, Jaehyun Bae, Ah Ra Kim, Kyu Hwan Lee, Se Hyun Kim, Myung Gwan Hahm, and Jaewook Nam. Electrohydrodynamic printing for scalable MoS 2 flake coating: application to gas sensing device. *Nanotechnology*, 27(43):435501, Oct 2016.

[251] Yousef Farraj, Michael Grouchko, and Shlomo Magdassi. Self-reduction of a copper complex mod ink for inkjet printing conductive patterns on plastics. *Chem. Commun.*, 51(9):1587–1590, 2015.

[252] Natsuki Komoda, Masaya Nogi, Katsuaki Suganuma, and Kanji Otsuka. Highly sensitive antenna using inkjet overprinting with particle-free conductive inks. *ACS Appl. Mater. Interfaces*, 4(11):5732–5736, Nov 2012.

[253] Yitzchak S. Rosen, Alexey Yakushenko, Andreas Offenhäusser, and Shlomo Magdassi. Self-reducing copper precursor inks and photonic additive yield conductive patterns under intense pulsed light. *ACS Omega*, 2(2):573–581, Feb 2017.

[254] Dong-Youn Shin, Minhwan Jung, and Sangki Chun. Resistivity transition mechanism of silver salts in the next generation conductive ink for a roll-to-roll printed film with a silver network. *J. Mater. Chem.*, 22(23):11755, 2012.

[255] Yu-Mo Chien, Florent Lefevre, Ishiang Shih, and Ricardo Izquierdo. A solution processed top emission oled with transparent carbon nanotube electrodes. *Nanotechnology*, 21(13):134020, Apr 2010.

[256] Taik-Min Lee, Jae-Ho Noh, Sun-Woo Kwak, Bongmin Kim, Jeongdai Jo, and Inyoung Kim. Design and fabrication of printed transparent electrode with silver mesh. *Microelectron. Eng.*, 98:556–560, Oct 2012.

[257] Thue T. Larsen-Olsen, Roar R. Søndergaard, Kion Norrman, Mikkel Jørgensen, and Frederik C. Krebs. All printed transparent electrodes through an electrical switching mechanism: a convincing alternative to indium-tin-oxide, silver and vacuum. *Energy Environ. Sci.*, 5(11):9467, 2012.

[258] Pak Heng Lau, Kuniharu Takei, Chuan Wang, Yeonkyeong Ju, Junseok Kim, Zhibin Yu, Toshitake Takahashi, Gyoujin Cho, and Ali Javey. Fully printed, high performance carbon nanotube thin-film transistors on flexible substrates. *Nano Lett.*, 13(8):3864–3869, Aug 2013.

[259] He Yan, Zhihua Chen, Yan Zheng, Christopher Newman, Jordan R. Quinn, Florian Dötz, Marcel Kastler, and Antonio Facchetti. A high-mobility electron-transporting polymer for printed transistors. *Nature*, 457(7230):679–686, Feb 2009.

[260] Xuan Cao, Haitian Chen, Xiaofei Gu, Bilu Liu, Wenli Wang, Yu Cao, Fanqi Wu, and Chongwu Zhou. Screen printing as a scalable and low-cost approach for rigid and flexible thin-film transistors using separated carbon nanotubes. *ACS Nano*, 8(12):12769–12776, Dec 2014.

[261] Jinsoo Noh, Minhoon Jung, Younsu Jung, Chisun Yeom, Myungho Pyo, and Gyoujin Cho. Key issues with printed flexible thin film transistors and their application in disposable RF sensors. *Proc. IEEE*, 103(4):554–566, Apr 2015.

[262] Alexandre Poulin, Xavier Aeby, Gilberto Siqueira, and Gustav Nyström. Versatile carbon-loaded shellac ink for disposable printed electronics. *Sci. Rep.*, 11(1):23784, Dec 2021.

[263] Yifei Wang, Holly Y. H. Kwok, Wending Pan, Yingguang Zhang, Huimin Zhang, Xu Lu, and Dennis Y. C. Leung. Printing Al-air batteries on paper for powering disposable printed electronics. *J. Power Sources*, 450:227685, Feb 2020.

[264] Rong Zhang and Yunfang Jia. A disposable printed liquid gate graphene field effect transistor for a salivary cortisol test. *ACS Sens.*, 6(8):3024–3031, Aug 2021.

[265] Suji Choi, Sang Ihn Han, Dongjun Jung, Hye Jin Hwang, Chaehong Lim, Soochan Bae, Ok Kyu Park, Cory M. Tschabrunn, Mincheol Lee, Sun Youn Bae, Ji Woong Yu, Ji Ho Ryu, Sang-Woo Lee, Kyungpyo Park, Peter M. Kang, Won Bo Lee, Reza Nezafat, Taeghwan Hyeon, and Dae-Hyeong Kim. Highly conductive, stretchable and biocompatible Ag–Au core–sheath nanowire composite for wearable and implantable bioelectronics. *Nat. Nanotechnol.*, 13(11):1048–1056, Nov 2018.

[266] Eun Roh, Byeong-Ung Hwang, Doil Kim, Bo-Yeong Kim, and Nae-Eung Lee. Stretchable, transparent, ultrasensitive, and patchable strain sensor for human–machine interfaces comprising a nanohybrid of carbon nanotubes and conductive elastomers. *ACS Nano*, 9(6):6252–6261, June 2015.

[267] Yong Lin Kong. Multiscale 3D printing of nanomaterial-based electronics and ingestible biomedical devices. In Hooman Mohseni, editor, *Integrated sensors for biological and neural sensing*, p. 25, Online Only, United States, Mar 2021. SPIE.

[268] Simiao Niu, Naoji Matsuhisa, Levent Beker, Jinxing Li, Sihong Wang, Jiechen Wang, Yuanwen Jiang, Xuzhou Yan, Youngjun Yun, William Burnett, Ada S. Y. Poon, Jeffery B.-H. Tok, Xiaodong Chen, and Zhenan Bao. A wireless body area sensor network based on stretchable passive tags. *Nat. Electron.*, 2(8):361–368, Aug 2019.

[269] Almudena Rivadeneyra, José Fernández-Salmerón, Manuel Agudo, Juan A. López-Villanueva, Luis Fermín Capitan-Vallvey, and Alberto J. Palma. Design and characterization of a low thermal drift capacitive humidity sensor by inkjet-printing. *Sens. Actuators B, Chem.*, 195:123–131, May 2014.

[270] Kenjiro Fukuda and Takao Someya. Recent progress in the development of printed thin-film transistors and circuits with high-resolution printing technology. *Adv. Mater.*, 29(25):1602736, July 2017.

[271] Zhuo Cao, E. Koukharenko, M. J. Tudor, R. N. Torah, and S. P. Beeby. Flexible screen printed thermoelectric generator with enhanced processes and materials. *Sens. Actuators Phys.*, 238:196–206, Feb 2016.

[272] Abhinav M. Gaikwad, Daniel A. Steingart, Tse Nga-Ng, David E. Schwartz, and Gregory L. Whiting. A flexible high potential printed battery for powering printed electronics. *Appl. Phys. Lett.*, 102(23):233302, June 2013.

[273] Pavlos Giannakou, Mateus G. Masteghin, Robert C. T. Slade, Steven J. Hinder, and Maxim Shkunov. Energy storage on demand: ultra-high-rate and high-energy-density inkjet-printed NiO micro-supercapacitors. *J. Mater. Chem. A*, 7(37):21496–21506, 2019.

[274] R. A. Street, T. N. Ng, D. E. Schwartz, G. L. Whiting, J. P. Lu, R. D. Bringans, and J. Veres. From printed transistors to printed smart systems. *Proc. IEEE*, 103(4):607–618, Apr 2015.

[275] Mi-Sun Lee, Kyongsoo Lee, So-Yun Kim, Heejoo Lee, Jihun Park, Kwang-Hyuk Choi, Han-Ki Kim, Dae-Gon Kim, Dae-Young Lee, SungWoo Nam, and Jang-Ung Park. High-performance, transparent, and stretchable electrodes using graphene–metal nanowire hybrid structures. *Nano Lett.*, 13(6):2814–2821, June 2013.

[276] Gerd Grau and Vivek Subramanian. Fully high-speed gravure printed, low-variability, high-performance organic polymer transistors with sub-5 V operation. *Adv. Electron. Mater.*, 2(4):1500328, Apr 2016.

[277] Sung Il Park, Daniel S. Brenner, Gunchul Shin, Clinton D. Morgan, Bryan A. Copits, Ha Uk Chung, Melanie Y. Pullen, Kyung Nim Noh, Steve Davidson, Soong Ju Oh, Jangyeol Yoon, Kyung-In Jang, Vijay K. Samineni, Megan Norman, Jose G. Grajales-Reyes, Sherri K. Vogt, Saranya S. Sundaram, Kellie M. Wilson, Jeong Sook Ha, Renxiao Xu, Taisong Pan, Tae-il Kim, Yonggang Huang, Michael C. Montana, Judith P. Golden, Michael R. Bruchas, Robert W. Gereau, and John A. Rogers. Soft, stretchable, fully implantable miniaturized optoelectronic systems for wireless optogenetics. *Nat. Biotechnol.*, 33(12):1280–1286, Dec 2015.

[278] V. Subramanian, J. M. J. Frechet, P. C. Chang, D. C. Huang, J. B. Lee, S. E. Molesa, A. R. Murphy, D. R. Redinger, and S. K. Volkman. Progress toward development of all-printed RFID tags: materials, processes, and devices. *Proc. IEEE*, 93(7):1330–1338, July 2005.

[279] Negar Sani, Mats Robertsson, Philip Cooper, Xin Wang, Magnus Svensson, Peter Andersson Ersman, Petronella Norberg, Marie Nilsson, David Nilsson, Xianjie Liu, Hjalmar Hesselbom, Laurent Akesso, Mats Fahlman, Xavier Crispin, Isak Engquist, Magnus Berggren, and Göran Gustafsson. All-printed diode operating at 1.6 GHz. *Proc. Natl. Acad. Sci.*, 111(33):11943–11948, Aug 2014.

[280] Tachibana Tomihisa, Katsuhiko Shirasawa, and Hidetaka Takato. Investigation of electrical shading loss of bifacial interdigitated-back-contact (IBC) crystalline silicon solar cells with screen-printed electrode. *Jpn. J. Appl. Phys.*, 59(11):116503, Nov 2020.

[281] Aurora Rizzo, Marco Mazzeo, Mariano Biasiucci, Roberto Cingolani, and Giuseppe Gigli. White electroluminescence from a microcontact-printing-deposited CdSe/Zns colloidal quantum-dot monolayer. *Small*, 4(12):2143–2147, Dec 2008.

[282] Vanessa Wood, Jonathan E. Halpert, Matthew J. Panzer, Moungi G. Bawendi, and Vladimir Bulović. Alternating current driven electroluminescence from ZnSe/ZnS:Mn/ZnS nanocrystals. *Nano Lett.*, 9(6):2367–2371, June 2009.

[283] M. Eritt, C. May, K. Leo, M. Toerker, and C. Radehaus. Oled manufacturing for large area lighting applications. *Thin Solid Films*, 518(11):3042–3045, Mar 2010.

[284] Yunlei Zhou, Chaoshan Zhao, Jiachen Wang, Yanzhen Li, Chenxin Li, Hangyu Zhu, Shuxuan Feng, Shitai Cao, and Desheng Kong. Stretchable high-permittivity nanocomposites for epidermal alternating-current electroluminescent displays. *ACS Mater. Lett.*, 1(5):511–518, Nov 2019.

[285] Daniel Janczak, Marcin Słoma, Grzegorz Wróblewski, Anna Mlożniak, and Małgorzata Jakubowska. Screen-printed resistive pressure sensors containing graphene nanoplatelets and carbon nanotubes. *Sensors*, 14(9):17304–17312, 2014.

[286] Marcin Słoma, Malgorzata Jakubowska, Andrzej Kolek, Krzysztof Mleczko, Piotr Ptak, Adam Witold Stadler, Zbigniew Zawiślak, and Anna Mlożniak. Investigations on printed elastic resistors containing carbon nanotubes. *J. Mater. Sci., Mater. Electron.*, 22(9):1321–1329, 2011.

[287] Dominika Ogończyk, Łukasz Tymecki, Iwona Wyżkiewicz, Robert Koncki, and Stanisław Głąb. Screen-printed disposable urease-based biosensors for inhibitive detection of heavy metal ions. *Sens. Actuators B, Chem.*, 106(1):450–454, Apr 2005.

[288] Rui Ren, Cuicui Leng, and Shusheng Zhang. A chronocoulometric DNA sensor based on screen-printed electrode doped with ionic liquid and polyaniline nanotubes. *Biosens. Bioelectron.*, 25(9):2089–2094, May 2010.

[289] H. Ago, K. Petritsch, M. S. P. Shaffer, A. H. Windle, and R. H. Friend. Composites of carbon nanotubes and conjugated polymers for photovoltaic devices. *Adv. Mater.*, 11(15):1281–1285, Oct 1999.

[290] Emmanuel Kymakis, Minas M. Stylianakis, George D. Spyropoulos, Emmanuel Stratakis, Emmanuel Koudoumas, and Costas Fotakis. Spin coated carbon nanotubes as the hole transport layer in organic photovoltaics. *Sol. Energy Mater. Sol. Cells*, 96:298–301, Jan 2012.

[291] M. Välimäki, P. Apilo, R. Po, E. Jansson, A. Bernardi, M. Ylikunnari, M. Vilkman, G. Corso, J. Puustinen, J. Tuominen, and J. Hast. R2R-printed inverted OPV modules—towards arbitrary patterned designs. *Nanoscale*, 7(21):9570–9580, 2015.

[292] Seigo Ito, Peter Chen, Pascal Comte, Mohammad Khaja Nazeeruddin, Paul Liska, Péter Péchy, and Michael Grätzel. Fabrication of screen-printing pastes from TiO2 powders for dye-sensitised solar cells. *Prog. Photovolt. Res. Appl.*, 15(7):603–612, Nov 2007.

[293] Junya Kubo, Yoshihiro Matsuo, Takahiro Wada, Akira Yamada, and Makoto Konagai. Fabrication of Cu(In,Ga)Se 2 films by a combination of mechanochemical synthesis, wet bead milling, and a screen printing/sintering process. *MRS Proc.*, 1165:1165–M05–13, 2009.

[294] Rommel Noufi and Ken Zweibel. High-efficiency CdTe and CIGS thin-film solar cells: highlights and challenges. In *2006 IEEE 4th world conference on photovoltaic energy conference*, pp. 317–320, Waikoloa, HI, 2006. IEEE.

[295] Sang Chul Lim, Seong Hyun Kim, Yong Suk Yang, Mi Young Lee, Su Yong Nam, and Jun Bin Ko. Organic thin-film transistor using high-resolution screen-printed electrodes. *Jpn. J. Appl. Phys.*, 48(8):081503, Aug 2009.

[296] T. Rueckes, K. Kim, E. Joselevich, G. Y. Tseng, C.-L. Cheung, and C. M. Lieber. Carbon nanotube-based nonvolatile random access memory for molecular computing. *Science*, 289:94–97, July 2000.

[297] Asrar Alam, Ghuzanfar Saeed, and Sooman Lim. Screen-printed activated carbon/silver nanocomposite electrode material for a high performance supercapacitor. *Mater. Lett.*, 273:127933, Aug 2020.

[298] Shingo Ohta, Shogo Komagata, Juntaro Seki, Tohru Saeki, Shinya Morishita, and Takahiko Asaoka. All-solid-state lithium ion battery using garnet-type oxide and Li3BO3 solid electrolytes fabricated by screen-printing. *J. Power Sources*, 238:53–56, Sept 2013.

[299] Saleem Khan, Leandro Lorenzelli, and Ravinder S. Dahiya. Technologies for printing sensors and electronics over large flexible substrates: a review. *IEEE Sens. J.*, 15(6):3164–3185, June 2015.

[300] Esa Kunnari, Jani Valkama, Marika Keskinen, and Pauliina Mansikkamäki. Environmental evaluation of new technology: printed electronics case study. *J. Clean. Prod.*, 17(9):791–799, June 2009.

[301] Ho anh-duc Nguyen, Nguyen Hoang, Kee-Hyun Shin, and Sangyoon Lee. Improvement of surface roughness and conductivity by calendering process for printed electronics. In *2011 8th international conference on ubiquitous robots and ambient intelligence (URAI)*, pp. 685–687, Incheon, Nov 2011. IEEE.

[302] Marcin Słoma. *Nanomateriały węglowe w technologii elektroniki drukowanej*. Oficyna Wydawnicza Politechniki Warszawskiej, Warszawa, 2017.

[303] Adobe Acrobat Team. Fast-forward—comparing a 1980s supercomputer to a modern smartphone. *Adobe Blog*, https://blog.adobe.com/en/publish/2022/11/08/fast-forward-comparing-1980s-supercomputer-to-modern-smartphone.

[304] Eric Macdonald, Rudy Salas, David Espalin, Mireya Perez, Efrain Aguilera, Dan Muse, and Ryan B. Wicker. 3D printing for the rapid prototyping of structural electronics. *IEEE Access*, 2:234–242, Dec 2014.

[305] Marcin Słoma. 3D printed electronics with nanomaterials. *Nanoscale*, 15(12):5623–5648, 2023.

[306] Michael Maher, Adrien Smith, and Jesse Margiotta. A synopsis of the defense advanced research projects agency (DARPA) investment in additive manufacture and what challenges remain. In Henry Helvajian, Alberto Piqué, Martin Wegener, and Bo Gu, editors, *Laser 3D Manufacturing*, Vol. 8970, p. 897002, San Francisco, California, United States, Mar 2014. DOI: https://doi.org/10.1117/12.2044725.

[307] Alberto Pique. *Direct-write technologies for rapid prototyping applications: sensors, electronics and integrated power sources*. Academic Press, San Diego, Calif. London, 2002.

[308] S. C. Mukhopadhyay and N. K. Suryadevara. *Internet of things: challenges and opportunities*, volume 9 of *Smart sensors, measurement and instrumentation*, pp. 1–17. Springer International Publishing, Cham, 2014.

[309] S. Stoukatch, F. Dupont, L. Seronveaux, D. Vandormael, and M. Kraft. Additive low temperature 3D printed electronic as enabling technology for IoT application. In *2017 IEEE 19th electronics packaging technology conference (EPTC)*, pp. 1–6, Singapore, Singapore, Dec 2017. IEEE.

[310] J. Hoerber, J. Glasschroeder, M. Pfeffer, J. Schilp, M. Zaeh, and J. Franke. Approaches for additive manufacturing of 3D electronic applications. *Proc. CIRP*, 17:806–811, 2014.

[311] Bernd Niese, Thomas Stichel, Philipp Amend, Uwe Urmoneit, Stephan Roth, and Michael Schmidt. Manufacturing of conductive circuits for embedding stereolithography by means of conductive adhesive and laser sintering. *Phys. Proc.*, 56:336–344, 2014.

[312] Bartłomiej Wałpuski and Marcin Słoma. Accelerated testing and reliability of FDM-based structural electronics. *Appl. Sci.*, 12(3):1110, Jan 2022.

[313] New Minnesota bridges super sensors scan tragedy before it strikes: first look (with video). *Pop. Mech.*, Oct 2009. https://www.popularmechanics.com/science/a2784/4258306/.

[314] Anne Eisenberg. Hot off the presses, conductive ink. *N.Y. Times*, June 2012.

[315] Suresh Babu Muttana, Rakesh Kumar Dey, and Arghya Sardar. Trends in lightweighting of BEVs: a review of strategies—part ii. *Auto Tech Rev*, 3(10):18–23, Oct 2014.

[316] Hengfeng Yan, Jimin Chen, and Jinyan Zhao. 3D-mid manufacturing via laser direct structuring with nanosecond laser pulses. *J. Polym. Eng.*, 36(9):957–962, Nov 2016.

[317] Teemu Alajoki, Matti Koponen, Markus Tuomikoski, Mikko Heikkinen, Antti Keranen, Kimmo Keranen, Jukka-Tapani Makinen, Janne Aikio, and Kari Ronka. Hybrid in-mould integration for novel electrical and optical features in 3D plastic products. In *2012 4th electronic system-integration technology conference*, pp. 1–6, Amsterdam, Netherlands, Sept 2012. IEEE.

[318] Michele Miliciani, Gian Maria Mendicino, and Mark T DeMeuse. In-mold electronics applications: the control of ink properties through the use of mixtures of electrically conductive pastes. *Flex. Print. Electron.*, 8(3):035011, Sept 2023.

[319] Sang Yoon Lee, Seong Hyun Jang, Hyun Kyung Lee, Jong Sun Kim, SangKug Lee, Ho Jun Song, Jae Woong Jung, Eui Sang Yoo, and Jun Choi. The development and investigation of highly stretchable conductive inks for 3-dimensional printed in-mold electronics. *Org. Electron.*, 85:105881, Oct 2020.

[320] Mona Bakr, Frederick Bossuyt, Jan Vanfleteren, and Yibo Su. Flexible microsystems using over-molding technology. *Procedia Manufacturing*, 52:26–31, 2020.

[321] Lindsey M. Bollig, Peter J. Hilpisch, Greg S. Mowry, and Brittany B. Nelson-Cheeseman. 3D printed magnetic polymer composite transformers. *J. Magn. Magn. Mater.*, 442:97–101, Nov 2017.

[322] Kamrul Hassan, Md Julker Nine, Tran Thanh Tung, Nathan Stanley, Pei Lay Yap, Hadi Rastin, Le Yu, and Dusan Losic. Functional inks and extrusion-based 3D printing of 2d materials: a review of current research and applications. *Nanoscale*, 12(37):19007–19042, 2020.

[323] Yong Lin Kong, Ian A. Tamargo, Hyoungsoo Kim, Blake N. Johnson, Maneesh K. Gupta, Tae-Wook Koh, Huai-An Chin, Daniel A. Steingart, Barry P. Rand, and Michael C. McAlpine. 3D printed quantum dot light-emitting diodes. *Nano Lett.*, 14(12):7017–7023, Dec 2014.

[324] Jennifer A. Lewis and Bok Y. Ahn. Three-dimensional printed electronics. *Nature*, 518(7537):42–43, Feb 2015.

[325] Ruitao Su, Sung Hyun Park, Xia Ouyang, Song Ih Ahn, and Michael C. McAlpine. 3D-printed flexible organic light-emitting diode displays. *Sci. Adv.*, 8(1):eabl8798, Jan 2022.

[326] G. R. Ruschau, S. Yoshikawa, and R. E. Newnham. Resistivities of conductive composites. *J. Appl. Phys.*, 72(3):953–959, Aug 1992.

[327] Mark Weber and Musa R. Kamal. Estimation of the volume resistivity of electrically conductive composites. *Polym. Compos.*, 18(6):711–725, Dec 1997.

[328] K. Gnanasekaran, T. Heijmans, S. Van Bennekom, H. Woldhuis, S. Wijnia, G. De With, and H. Friedrich. 3D printing of CNT- and graphene-based conductive polymer nanocomposites by fused deposition modeling. *Appl. Mater. Today*, 9:21–28, Dec 2017.

[329] Simon J. Leigh, Robert J. Bradley, Christopher P. Purssell, Duncan R. Billson, and David A. Hutchins. A simple, low-cost conductive composite material for 3D printing of electronic sensors. *PLoS ONE*, 7(11):e49365, Nov 2012.

[330] Joseph T. Muth, Daniel M. Vogt, Ryan L. Truby, Yiğit Mengüç, David B. Kolesky, Robert J. Wood, and Jennifer A. Lewis. Embedded 3D printing of strain sensors within highly stretchable elastomers. *Adv. Mater.*, 26(36):6307–6312, Sept 2014.

[331] Bartłomiej Podsiadły, Andrzej Skalski, and Marcin Słoma. Conductive ABS/Ni composite filaments for fused deposition modeling of structural electronics. In *Mechatronics 2019: recent advances towards industry 4.0*, volume 1044, pp. 62–70, Cham, 2020. Springer International Publishing.

[332] Bartłomiej J. Podsiadły, Andrzej Skalski, and Marcin Słoma. Mechanical and thermal properties of ABS/iron composite for fused deposition modeling. In Ryszard S. Romaniuk and Maciej Linczuk, editors, *Photonics applications in astronomy, communications, industry, and high-energy physics experiments 2019*, p. 135, Wilga, Poland, Nov 2019. SPIE.

[333] Scott Kirkpatrick. Percolation and conduction. *Rev. Mod. Phys.*, 45(4):574–588, Oct 1973.

[334] Neil White. *Thick films*, p. 1. Springer handbooks. Springer International Publishing, Cham, 2017.

[335] Alessandro Chiolerio, Krishna Rajan, Ignazio Roppolo, Annalisa Chiappone, Sergio Bocchini, and Denis Perrone. Silver nanoparticle ink technology: state of the art. *Nanotechnol. Sci. Appl.*, p. 1, Jan 2016.

[336] Kyoung-Sik Moon, Hai Dong, Radenka Maric, Suresh Pothukuchi, Andrew Hunt, Yi Li, and C. P. Wong. Thermal behavior of silver nanoparticles for low-temperature interconnect applications. *J. Electron. Mater.*, 34(2):168–175, Feb 2005.

[337] Małgorzata Jakubowska, Marcin Słoma, and Anna Młożniak. Printed transparent electrodes containing carbon nanotubes for elastic circuits applications with enhanced electrical durability under severe conditions. *Mater. Sci. Eng. B*, 176(4):358–362, Mar 2011.

[338] Pawel Kopyt, Bartlomiej Salski, Marzena Olszewska-Placha, Daniel Janczak, Marcin Słoma, Tomasz Kurkus, Malgorzata Jakubowska, and Wojciech Gwarek. Graphene-based dipole antenna for a UHF RFID tag. *IEEE Trans. Antennas Propag.*, 64(7):2862–2868, July 2016.

[339] Marcin Słoma, Grzegorz Wróblewski, Daniel Janczak, and Malgorzata Jakubowska. Transparent electrodes with nanotubes and graphene for printed optoelectronic applications. *J. Nanomater.*, 2014:17, Jan 2014.

[340] Yejung Choi, Kwang-dong Seong, and Yuanzhe Piao. Metal-organic decomposition ink for printed electronics. *Adv. Mater. Interfaces*, 6(20):1901002, Oct 2019.

[341] Gerard Cummins, Robert Kay, Jonathan Terry, Marc P. Y. Desmulliez, and Anthony J. Walton. Optimization and characterization of drop-on-demand inkjet printing process for platinum organometallic inks. In *2011 IEEE 13th electronics packaging technology conference*, pp. 256–261, Singapore, Singapore, Dec 2011. IEEE.

[342] Calvin J. Curtis, Alexander Miedaner, Marinus Franciscus Antonius Maria van Hest, and David S. Ginley. Printing aluminum films and patterned contacts using organometallic precursor inks, Aug 2010. https://patents.google.com/patent/US20100209594A1/en, US20100209594A1.

[343] Hye Moon Lee, Si-Young Choi, Kyung Tae Kim, Jung-Yeul Yun, Dae Soo Jung, Seung Bin Park, and Jongwook Park. A novel solution-stamping process for preparation of a highly conductive aluminum thin film. *Adv. Mater.*, 23(46):5524–5528, Dec 2011.

[344] Neha Thakur and Hari Murthy. Nickel-based inks for inkjet printing: a review on latest trends. *Am. J. Mater. Sci.*, 11(1):20–35, 2021.

[345] Yue Dong, Xiaodong Li, Shaohong Liu, Qi Zhu, Ji-Guang Li, and Xudong Sun. Facile synthesis of high silver content mod ink by using silver oxalate precursor for inkjet printing applications. *Thin Solid Films*, 589:381–387, Aug 2015.

[346] Pengbing Zhao, Jin Huang, Jinzheng Nan, Dachuan Liu, and Fanbo Meng. Laser sintering process optimization of microstrip antenna fabricated by inkjet printing with silver-based mod ink. *J. Mater. Process. Technol.*, 275:116347, Jan 2020.

[347] H. M. Nur, J. H. Song, J. R. G. Evans, and M. J. Edirisinghe. Ink-jet printing of gold conductive tracks. *J. Mater. Sci., Mater. Electron.*, 13(4):213–219, 2002.

[348] Claudia Schoner, André Tuchscherer, Thomas Blaudeck, Stephan F. Jahn, Reinhard R. Baumann, and Heinrich Lang. Particle-free gold metal–organic decomposition ink for inkjet printing of gold structures. *Thin Solid Films*, 531:147–151, Mar 2013.

[349] Soonchul Kang, Kazuya Tasaka, Ji Ha Lee, and Akihiro Yabuki. Self-reducible copper complex inks with two amines for copper conductive films via calcination below 100 °C. *Chem. Phys. Lett.*, 763:138248, Jan 2021.

[350] Jeonghyeon Lee, Byoungyoon Lee, Sooncheol Jeong, Yoonhyun Kim, and Myeongkyu Lee. Microstructure and electrical property of laser-sintered Cu complex ink. *Appl. Surf. Sci.*, 307:42–45, July 2014.

[351] Dong-Hun Shin, Seunghee Woo, Hyesuk Yem, Minjeong Cha, Sanghun Cho, Mingyu Kang, Sooncheol Jeong, Yoonhyun Kim, Kyungtae Kang, and Yuanzhe Piao. A self-reducible and alcohol-soluble copper-based metal–organic decomposition ink for printed electronics. *ACS Appl. Mater. Interfaces*, 6(5):3312–3319, Mar 2014.

[352] Wen Xu and Tao Wang. Synergetic effect of blended alkylamines for copper complex ink to form conductive copper films. *Langmuir*, 33(1):82–90, Jan 2017.

[353] Guo Liang Goh, Haining Zhang, Tzyy Haur Chong, and Wai Yee Yeong. 3D printing of multilayered and multimaterial electronics: a review. *Adv. Electron. Mater.*, 7(10):2100445, Oct 2021.

[354] Ethan B. Secor, Pradyumna L. Prabhumirashi, Kanan Puntambekar, Michael L. Geier, and Mark C. Hersam. Inkjet printing of high conductivity, flexible graphene patterns. *J. Phys. Chem. Lett.*, 4(8):1347–1351, Apr 2013.

[355] Jong-Hyun Ahn, Hoon-Sik Kim, Keon Jae Lee, Seokwoo Jeon, Seong Jun Kang, Yugang Sun, Ralph G. Nuzzo, and John A. Rogers. Heterogeneous three-dimensional electronics by use of printed semiconductor nanomaterials. *Science*, 314(5806):1754–1757, Dec 2006.

[356] Suresh Kumar Garlapati, Mitta Divya, Ben Breitung, Robert Kruk, Horst Hahn, and Subho Dasgupta. Printed electronics based on inorganic semiconductors: from processes and materials to devices. *Adv. Mater.*, 30(40):1707600, Oct 2018.

[357] K. I. Bolotin, K. J. Sikes, Z. Jiang, M. Klima, G. Fudenberg, J. Hone, P. Kim, and H. L. Stormer. Ultrahigh electron mobility in suspended graphene. *Solid State Commun.*, 146(9–10):351–355, June 2008.

[358] Jianwen Zhao, Yulong Gao, Jian Lin, Zheng Chen, and Zheng Cui. Printed thin-film transistors with functionalized single-walled carbon nanotube inks. *J. Mater. Chem.*, 22(5):2051–2056, 2012.

[359] Meikang Han, Danzhen Zhang, Christopher E. Shuck, Bernard McBride, Teng Zhang, Ruocun Wang, Kateryna Shevchuk, and Yury Gogotsi. Electrochemically modulated interaction of MXenes with microwaves. *Nat. Nanotechnol.*, Jan 2023.

[360] Dmitri V. Talapin and Jonathan Steckel. Quantum dot light-emitting devices. *Mater. Res. Soc. Bull.*, 38(9):685–691, Sept 2013.

[361] Seungjun Chung, Kyungjune Cho, and Takhee Lee. Recent progress in inkjet-printed thin-film transistors. *Adv. Sci.*, 6(6):1801445, Mar 2019.

[362] Daniel Corzo, Diego Rosas-Villalva, Amruth C, Guillermo Tostado-Blázquez, Emily Bezerra Alexandre, Luis Huerta Hernandez, Jianhua Han, Han Xu, Maxime Babics, Stefaan De Wolf, and Derya Baran. High-performing organic electronics using terpene green solvents from renewable feedstocks. *Nat. Energy*, Dec 2022.

[363] Jimin Kwon, Yasunori Takeda, Kenjiro Fukuda, Kilwon Cho, Shizuo Tokito, and Sungjune Jung. Three-dimensional, inkjet-printed organic transistors and integrated circuits with 100 % yield, high uniformity, and long-term stability. *ACS Nano*, 10(11):10324–10330, Nov 2016.

[364] Yugeng Wen and Yunqi Liu. Recent progress in n-channel organic thin-film transistors. *Adv. Mater.*, 22(12):1331–1345, Mar 2010.

[365] Jeesoo Seok, Tae Joo Shin, Sungmin Park, Changsoon Cho, Jung-Yong Lee, Du Yeol Ryu, Myung Hwa Kim, and Kyungkon Kim. Efficient organic photovoltaics utilizing nanoscale heterojunctions in sequentially deposited polymer/fullerene bilayer. *Sci. Rep.*, 5:8373, Feb 2015.

[366] Yu-Wei Su, Shang-Che Lan, and Kung-Hwa Wei. Organic photovoltaics. *Mater. Today*, 15(12):554–562, Dec 2012.

[367] Matteo Massetti, Silan Zhang, Padinhare Cholakkal Harikesh, Bernhard Burtscher, Chiara Diacci, Daniel T. Simon, Xianjie Liu, Mats Fahlman, Deyu Tu, Magnus Berggren, and Simone Fabiano. Fully 3D-printed organic electrochemical transistors. *Npj Flex. Electron.*, 7(1):11, Mar 2023.

[368] Yotsarayuth Seekaew, Shongpun Lokavee, Ditsayut Phokharatkul, Anurat Wisitsoraat, Teerakiat Kerdcharoen, and Chatchawal Wongchoosuk. Low-cost and flexible printed graphene–PEDOT:PSS gas sensor for ammonia detection. *Org. Electron.*, 15(11):2971–2981, Nov 2014.

[369] Hugo Bronstein, Christian B. Nielsen, Bob C. Schroeder, and Iain McCulloch. The role of chemical design in the performance of organic semiconductors. *Nat. Rev. Chem.*, 4(2):66–77, Jan 2020.

[370] Wolfgang Brütting, editor. *Physics of organic semiconductors*. Wiley, 1st edition, May 2005.

[371] Florian Machui, Steven Abbott, David Waller, Markus Koppe, and Christoph J. Brabec. Determination of solubility parameters for organic semiconductor formulations: determination of solubility parameters for organic semiconductor …. *Macromol. Chem. Phys.*, 212(19):2159–2165, Oct 2011.

[372] Solmaz Torabi, Fatemeh Jahani, Ineke Van Severen, Catherine Kanimozhi, Satish Patil, Remco W. A. Havenith, Ryan C. Chiechi, Laurence Lutsen, Dirk J. M. Vanderzande, Thomas J. Cleij, Jan C. Hummelen, and L. Jan Anton Koster. Strategy for enhancing the dielectric constant of organic semiconductors without sacrificing charge carrier mobility and solubility. *Adv. Funct. Mater.*, 25(1):150–157, Jan 2015.

[373] J. Varghese and M. T. Sebastian. *Dielectric inks*, pp. 457–480. Wiley, 1st edition, Apr 2017.

[374] Zhongbo Zhang, Jifu Zheng, Kasun Premasiri, Man-Hin Kwok, Qiong Li, Ruipeng Li, Suobo Zhang, Morton H. Litt, Xuan P. A. Gao, and Lei Zhu. High-k polymers of intrinsic microporosity: a new class of high temperature and low loss dielectrics for printed electronics. *Mater. Horiz.*, 7(2):592–597, 2020.

[375] Stefan Christian Endres, Lucio Colombi Ciacchi, and Lutz Mädler. A review of contact force models between nanoparticles in agglomerates, aggregates, and films. *J. Aerosol Sci.*, 153:105719, Mar 2021.

[376] Graham A. Rance, Dan H. Marsh, Stephen J. Bourne, Thomas J. Reade, and Andrei N. Khlobystov. van der Waals interactions between nanotubes and nanoparticles for controlled assembly of composite nanostructures. *ACS Nano*, 4(8):4920–4928, Aug 2010.

[377] Andrzej Galeski. Strength and toughness of crystalline polymer systems. *Prog. Polym. Sci.*, 28(12):1643–1699, Dec 2003.

[378] Jang-Kyo Kim and Y. W. Mai. *Engineered interfaces in fiber reinforced composites*. Elsevier Sciences, Amsterdam; New York, 1st edition, 1998.

[379] A Hodzic, Z. H Stachurski, and J. K Kim. Nano-indentation of polymer–glass interfaces part i. Experimental and mechanical analysis. *Polymer*, 41(18):6895–6905, Aug 2000.

[380] Jang-Kyo Kim, Man-Lung Sham, and Jingshen Wu. Nanoscale characterisation of interphase in silane treated glass fibre composites. *Composites, Part A, Appl. Sci. Manuf.*, 32(5):607–618, May 2001.

[381] Erik T Thostenson, Zhifeng Ren, and Tsu-Wei Chou. Advances in the science and technology of carbon nanotubes and their composites: a review. *Compos. Sci. Technol.*, 61(13):1899–1912, Oct 2001.

[382] J. N. Coleman, U. Khan, and Y. K. Gun'ko. Mechanical reinforcement of polymers using carbon nanotubes. *Adv. Mater.*, 18(6):689–706, Mar 2006.

[383] Shengfeng Cheng and Gary S. Grest. Dispersing nanoparticles in a polymer film via solvent evaporation. *ACS Macro Lett.*, 5(6):694–698, June 2016.

[384] Arnaldo D. Valino, John Ryan C. Dizon, Alejandro H. Espera, Qiyi Chen, Jamie Messman, and Rigoberto C. Advincula. Advances in 3D printing of thermoplastic polymer composites and nanocomposites. *Prog. Polym. Sci.*, 98:101162, Nov 2019.

[385] JrH Du, Jinbo Bai, HrM Cheng, et al.The present status and key problems of carbon nanotube based polymer composites. *eXPRESS Polym. Lett.*, 1(5):253–273, 2007.

[386] Nadia Grossiord, Joachim Loos, Oren Regev, and Cor E. Koning. Toolbox for dispersing carbon nanotubes into polymers to get conductive nanocomposites. *Chem. Mater.*, 18(5):1089–1099, Mar 2006.

[387] Mohammad Moniruzzaman and Karen I. Winey. Polymer nanocomposites containing carbon nanotubes. *Macromolecules*, 39(16):5194–5205, Aug 2006.

[388] Achilleas Sesis, Mark Hodnett, Gianluca Memoli, Andrew J. Wain, Izabela Jurewicz, Alan B. Dalton, J. David Carey, and Gareth Hinds. Influence of acoustic cavitation on the controlled ultrasonic dispersion of carbon nanotubes. *J. Phys. Chem. B*, p. 131125074637001, Nov 2013.

[389] Jiajie Liang, Yi Huang, Long Zhang, Yan Wang, Yanfeng Ma, Tianyin Guo, and Yongsheng Chen. Molecular-level dispersion of graphene into poly(vinyl alcohol) and effective reinforcement of their nanocomposites. *Adv. Funct. Mater.*, 19(14):2297–2302, July 2009.

[390] Yuxi Xu, Wenjing Hong, Hua Bai, Chun Li, and Gaoquan Shi. Strong and ductile poly(vinyl alcohol)/graphene oxide composite films with a layered structure. *Carbon*, 47(15):3538–3543, Dec 2009.

[391] T. Ramanathan, S. Stankovich, D. A. Dikin, H. Liu, H. Shen, S. T. Nguyen, and L. C. Brinson. Graphitic nanofillers in PMMA nanocomposites—an investigation of particle size and dispersion and their influence on nanocomposite properties. *J. Polym. Sci., Part B, Polym. Phys.*, 45(15):2097–2112, Aug 2007.

[392] Murielle Cochet, Wolfgang K. Maser, Ana M. Benito, M. Alicia Callejas, M. Teresa Martínez, Jean-Michel Benoit, Joachim Schreiber, and Olivier Chauvet. Synthesis of a new polyaniline/nanotube composite: "in-situ" polymerisation and charge transfer through site-selective interaction. *Chem. Commun.*, (16):1450–1451, 2001.

[393] Qiong Wu, Yuxi Xu, Zhiyi Yao, Anran Liu, and Gaoquan Shi. Supercapacitors based on flexible graphene/polyaniline nanofiber composite films. *ACS Nano*, 4(4):1963–1970, Apr 2010.

[394] Dongyu Cai and Mo Song. A simple route to enhance the interface between graphite oxide nanoplatelets and a semi-crystalline polymer for stress transfer. *Nanotechnology*, 20(31):315708, Aug 2009.

[395] Dongyu Cai, Kamal Yusoh, and Mo Song. The mechanical properties and morphology of a graphite oxide nanoplatelet/polyurethane composite. *Nanotechnology*, 20(8):085712, Feb 2009.

[396] Jiajie Liang, Yanfei Xu, Yi Huang, Long Zhang, Yan Wang, Yanfeng Ma, Feifei Li, Tianying Guo, and Yongsheng Chen. Infrared-triggered actuators from graphene-based nanocomposites. *J. Phys. Chem. C*, 113(22):9921–9927, June 2009.

[397]   Marcin Słoma. *Opracowanie technologii i badania właściwości kompozytów polimerowych z nanorurkami węglowymi i ich zastosowania.* Rozprawa doktorska, Politechnika Warszawska, Feb 2011.

[398]   K. L. Lu, R. M. Lago, Y. K. Chen, M. L. H. Green, P. J. F. Harris, and S. C. Tsang. *Mechanical damage of carbon nanotubes by ultrasound*, volume 34 of *Carbon*. Elsevier, Kidlington, ROYAUME-UNI, 1996.

[399]   Kingsuk Mukhopadhyay, Dwivedi Chandra Dhar, and Mathur Gyanesh Narayan. *Conversion of carbon nanotubes to carbon nanofibers by sonication*, volume 40 of *Carbon*. Elsevier, Kidlington, ROYAUME-UNI, 2002.

[400]   F. H. Gojny, M. H. G. Wichmann, U. Köpke, B. Fiedler, and K. Schulte. Carbon nanotube-reinforced epoxy-composites: enhanced stiffness and fracture toughness at low nanotube content. *Compos. Sci. Technol.*, 64(15):2363–2371, Nov 2004.

[401]   Erik T. Thostenson and Tsu-Wei Chou. Processing-structure-multi-functional property relationship in carbon nanotube/epoxy composites. *Carbon*, 44(14):3022–3029, Nov 2006.

[402]   EXAKT company. www.exakt.com.

[403]   J. Sandler, M. S. P. Shaffer, T. Prasse, W. Bauhofer, K. Schulte, and A. H. Windle. Development of a dispersion process for carbon nanotubes in an epoxy matrix and the resulting electrical properties. *Polymer*, 40(21):5967–5971, Oct 1999.

[404]   J. Li, P. C. Ma, W. S. Chow, C. K. To, B. Z. Tang, and J.-K. Kim. Correlations between percolation threshold, dispersion state, and aspect ratio of carbon nanotubes. *Adv. Funct. Mater.*, 17(16):3207–3215, Nov 2007.

[405]   C. F. Schmid and D. J. Klingenberg. Mechanical flocculation in flowing fiber suspensions. *Phys. Rev. Lett.*, 84(2):290–293, Jan 2000.

[406]   K. Awasthi. Ball-milled carbon and hydrogen storage. *Int. J. Hydrog. Energy*, 27(4):425–432, Apr 2002.

[407]   B. Gao, C. Bower, J. D. Lorentzen, L. Fleming, A. Kleinhammes, X. P. Tang, L. E. McNeil, Y. Wu, and O. Zhou. Enhanced saturation lithium composition in ball-milled single-walled carbon nanotubes. *Chem. Phys. Lett.*, 327(1–2):69–75, Sept 2000.

[408]   Y. A. Kim, T. Hayashi, Y. Fukai, M. Endo, T. Yanagisawa, and M. S. Dresselhaus. Effect of ball milling on morphology of cup-stacked carbon nanotubes. *Chem. Phys. Lett.*, 355(3–4):279–284, Apr 2002.

[409]   Y. B. Li, B. Q. Wei, J. Liang, Q. Yu, and D. H. Wu. Transformation of carbon nanotubes to nanoparticles by ball milling process. *Carbon*, 37(3):493–497, Feb 1999.

[410]   Peng Cheng Ma, Ben Zhong Tang, and Jang-Kyo Kim. Conversion of semiconducting behavior of carbon nanotubes using ball milling. *Chem. Phys. Lett.*, 458(1–3):166–169, June 2008.

[411]   Peng Cheng Ma, Sheng Qi Wang, Jang-Kyo Kim, and Ben Zhong Tang. In-situ amino functionalization of carbon nanotubes using ball milling. *J. Nanosci. Nanotechnol.*, 9(2):749–753, Feb 2009.

[412]   Yanfei Xu, Matthias Georg Schwab, Andrew James Strudwick, Ingolf Hennig, Xinliang Feng, Zhongshuai Wu, and Klaus Müllen. Screen-printable thin film supercapacitor device utilizing graphene/polyaniline inks. *Adv. Energy Mater.*, 3(8):1035–1040, Aug 2013.

[413]   Tobias Villmow, Petra Pötschke, Sven Pegel, Liane Häussler, and Bernd Kretzschmar. Influence of twin-screw extrusion conditions on the dispersion of multi-walled carbon nanotubes in a poly(lactic acid) matrix. *Polymer*, 49(16):3500–3509, July 2008.

[414]   Dietrich Stauffer and Ammon Aharony. *Introduction to percolation theory: second edition.* CRC Press, Dec 2018.

[415]   S. R. Broadbent and J. M. Hammersley. Percolation processes: I. Crystals and mazes. *Math. Proc. Camb. Philos. Soc.*, 53(3):629–641, July 1957.

[416]   Hong Wei Tan, Jia An, Chee Kai Chua, and Tuan Tran. Metallic nanoparticle inks for 3D printing of electronics. *Adv. Electron. Mater.*, 5(5):1800831, May 2019.

[417]   Cornell University. Loudspeaker is first-ever 3-d-printed consumer electronic. *ScienceDaily*, Dec 2013. www.sciencedaily.com/releases/2013/12/131216142224.htm.

[418]   Jeremy Hsu. First 3-d printed loudspeaker hints at future of consumer electronics, *IEEE Spectr.*, Dec 2013. http://spectrum.ieee.org/tech-talk/consumer-electronics/gadgets/first-3d-printed-loudspeaker-hints-at-future-of-consumer-electronics.

[419] Bingheng Lu, Hongbo Lan, and Hongzhong Liu. Additive manufacturing frontier: 3D printing electronics. *Opto-Electron. Adv.*, 1(1):17000401–17000410, 2018.

[420] S. Brett Walker and Jennifer A. Lewis. Reactive silver inks for patterning high-conductivity features at mild temperatures. *J. Am. Chem. Soc.*, 134(3):1419–1421, Jan 2012.

[421] F. Castles, D. Isakov, A. Lui, Q. Lei, C. E. J. Dancer, Y. Wang, J. M. Janurudin, S. C. Speller, C. R. M. Grovenor, and P. S. Grant. Microwave dielectric characterisation of 3D-printed BaTiO3/ABS polymer composites. *Sci. Rep.*, 6(1):22714, Sept 2016.

[422] Sung-Yueh Wu, Chen Yang, Wensyang Hsu, and Liwei Lin. 3D-printed microelectronics for integrated circuitry and passive wireless sensors. *Microsyst. Nanoeng.*, 1(1):15013, Dec 2015.

[423] Sylvia Castillo, Dan Muse, Frank Medina, Eric MacDonald, and Ryan Wicker. Electronics integration in conformal substrates fabricated with additive layered manufacturing. In *Proceedings of the 20th annual solid freeform fabrication symposium*, Sept 2009.

[424] Amit Joe Lopes, Eric MacDonald, and Ryan B. Wicker. Integrating stereolithography and direct print technologies for 3D structural electronics fabrication. *Rapid Prototyping J.*, 18(2):129–143, Mar 2012.

[425] David Espalin, Danny W. Muse, Eric MacDonald, and Ryan B. Wicker. 3D printing multifunctionality: structures with electronics. *Int. J. Adv. Manuf. Technol.*, 72(5–8):963–978, May 2014.

[426] Eric MacDonald and Ryan Wicker. Multiprocess 3D printing for increasing component functionality. *Science*, 353(6307):aaf2093, Sept 2016.

[427] Xi Zhang, Tong Ge, and Joseph S. Chang. Fully-additive printed electronics: transistor model, process variation and fundamental circuit designs. *Org. Electron.*, 26:371–379, Nov 2015.

[428] Matthew Dyson. *3D electronics/additive electronics 2022–2032*. https://www.idtechex.com/en/research-report/3d-electronics-additive-electronics-2022-2032/860. ISBN 978-1-913899-93-6.

[429] Chee Kai Chua, Wai Yee Yeong, Hong Yee Low, Tuan Tran, and Hong Wei Tan. *3D printing and additive manufacturing of electronics: principles and applications*. World Scientific, May 2021.

[430] Kyuyoung Kim, Jaeho Park, Ji-hoon Suh, Minseong Kim, Yongrok Jeong, and Inkyu Park. 3D printing of multiaxial force sensors using carbon nanotube (CNT)/thermoplastic polyurethane (TPU) filaments. *Sens. Actuators Phys.*, 263:493–500, Aug 2017.

[431] Shaohong Shi, Yinghong Chen, Jingjing Jing, and Lu Yang. Preparation and 3D-printing of highly conductive polylactic acid/carbon nanotube nanocomposites via local enrichment strategy. *RSC Adv.*, 9(51):29980–29986, 2019.

[432] Sithiprumnea Dul, Luca Fambri, and Alessandro Pegoretti. Fused deposition modelling with ABS–graphene nanocomposites. *Composites, Part A, Appl. Sci. Manuf.*, 85:181–191, June 2016.

[433] Xiaojun Wei, Dong Li, Wei Jiang, Zheming Gu, Xiaojuan Wang, Zengxing Zhang, and Zhengzong Sun. 3D printable graphene composite. *Sci. Rep.*, 5(1):11181, Sept 2015.

[434] Lei Lei, Zhengjun Yao, Jintang Zhou, Bo Wei, and Huiyuan Fan. 3D printing of carbon black/polypropylene composites with excellent microwave absorption performance. *Compos. Sci. Technol.*, 200:108479, Nov 2020.

[435] Kun Fu, Yonggang Yao, Jiaqi Dai, and Liangbing Hu. Progress in 3D printing of carbon materials for energy-related applications. *Adv. Mater.*, 29(9):1603486, Mar 2017.

[436] Nathan Lazarus and Sarah S Bedair. Creating 3D printed sensor systems with conductive composites. *Smart Mater. Struct.*, 30(1):015020, Jan 2021.

[437] A. S. Luyt, J. A. Molefi, and H. Krump. Thermal, mechanical and electrical properties of copper powder filled low-density and linear low-density polyethylene composites. *Polym. Degrad. Stab.*, 91(7):1629–1636, July 2006.

[438] Ye. P. Mamunya, V. V. Davydenko, P. Pissis, and E. V. Lebedev. Electrical and thermal conductivity of polymers filled with metal powders. *Eur. Polym. J.*, 38(9):1887–1897, Sept 2002.

[439] Kambiz Chizari, Mohamed Amine Daoud, Anil Raj Ravindran, and Daniel Therriault. 3D printing of highly conductive nanocomposites for the functional optimization of liquid sensors. *Small*, 12(44):6076–6082, Nov 2016.

[440] Gustavo Gonzalez, Annalisa Chiappone, Ignazio Roppolo, Erika Fantino, Valentina Bertana, Francesco Perrucci, Luciano Scaltrito, Fabrizio Pirri, and Marco Sangermano. Development of 3D printable formulations containing CNT with enhanced electrical properties. *Polymer*, 109:246–253, Jan 2017.

[441] Gregorio de la Osa, Domingo Pérez-Coll, Pilar Miranzo, María Isabel Osendi, and Manuel Belmonte. Printing of graphene nanoplatelets into highly electrically conductive three-dimensional porous macrostructures. *Chem. Mater.*, 28(17):6321–6328, Sept 2016.

[442] Sepidar Sayyar, Sanjeev Gambhir, Johnson Chung, David L. Officer, and Gordon G. Wallace. 3D printable conducting hydrogels containing chemically converted graphene. *Nanoscale*, 9(5):2038–2050, 2017.

[443] José F. Salmerón, Francisco Molina-Lopez, Danick Briand, Jason J. Ruan, Almudena Rivadeneyra, Miguel A. Carvajal, L. F. Capitán-Vallvey, Nico F. De Rooij, and Alberto J. Palma. Properties and printability of inkjet and screen-printed silver patterns for RFID antennas. *J. Electron. Mater.*, 43(2):604–617, Feb 2014.

[444] Bok Yeop Ahn, David J. Lorang, and Jennifer A. Lewis. Transparent conductive grids via direct writing of silver nanoparticle inks. *Nanoscale*, 3(7):2700, 2011.

[445] Krishnamraju Ankireddy, Swathi Vunnam, Jon Kellar, and William Cross. Highly conductive short chain carboxylic acid encapsulated silver nanoparticle based inks for direct write technology applications. *J. Mater. Chem. C*, 1(3):572–579, 2013.

[446] Yejin Jo, Ju Young Kim, Sungmook Jung, Bok Yeop Ahn, Jennifer A. Lewis, Youngmin Choi, and Sunho Jeong. 3D polymer objects with electronic components interconnected via conformally printed electrodes. *Nanoscale*, 9(39):14798–14803, 2017.

[447] Bartłomiej Wałpuski and Marcin Słoma. Additive manufacturing of electronics from silver nanopowders sintered on 3D printed low-temperature substrates. *Adv. Eng. Mater.*, 23(4):2001085, Apr 2021.

[448] Wen-Tao Cao, Chang Ma, Dong-Sheng Mao, Juan Zhang, Ming-Guo Ma, and Feng Chen. MXene-reinforced cellulose nanofibril inks for 3D-printed smart fibres and textiles. *Adv. Funct. Mater.*, 29(51):1905898, Dec 2019.

[449] Hadi Rastin, Bingyang Zhang, Arash Mazinani, Kamrul Hassan, Jingxiu Bi, Tran Thanh Tung, and Dusan Losic. 3D bioprinting of cell-laden electroconductive MXene nanocomposite bioinks. *Nanoscale*, 12(30):16069–16080, 2020.

[450] Chuanfang Zhang, Lorcan McKeon, Matthias P. Kremer, Sang-Hoon Park, Oskar Ronan, Andrés Seral-Ascaso, Sebastian Barwich, Cormac O Coileain, Niall McEvoy, Hannah C. Nerl, Babak Anasori, Jonathan N. Coleman, Yury Gogotsi, and Valeria Nicolosi. Additive-free MXene inks and direct printing of micro-supercapacitors. *Nat. Commun.*, 10(1):1795, Dec 2019.

[451] Teng-Bo Ma, Hao Ma, Kun-Peng Ruan, Xue-Tao Shi, Hua Qiu, Sheng-Yuan Gao, and Jun-Wei Gu. Thermally conductive poly(lactic acid) composites with superior electromagnetic shielding performances via 3D printing technology. *Chin. J. Polym. Sci.*, 40(3):248–255, Mar 2022.

[452] Xinyu Wu, Tingxiang Tu, Yang Dai, Pingping Tang, Yu Zhang, Zhiming Deng, Lulu Li, Hao-Bin Zhang, and Zhong-Zhen Yu. Direct ink writing of highly conductive MXene frames for tunable electromagnetic interference shielding and electromagnetic wave-induced thermochromism. *Nano-Micro Lett.*, 13(1):148, Dec 2021.

[453] Hyunwoo Yuk, Baoyang Lu, Shen Lin, Kai Qu, Jingkun Xu, Jianhong Luo, and Xuanhe Zhao. 3D printing of conducting polymers. *Nat. Commun.*, 11(1):1604, Mar 2020.

[454] Muhammad Wajahat, Jung Hyun Kim, Jinhyuck Ahn, Sanghyeon Lee, Jongcheon Bae, Jaeyeon Pyo, and Seung Kwon Seol. 3D printing of Fe3O4 functionalized graphene-polymer (FGP) composite microarchitectures. *Carbon*, 167:278–284, Oct 2020.

[455] Abhishek Gannarapu and Bulent Arda Gozen. Freeze-printing of liquid metal alloys for manufacturing of 3D, conductive, and flexible networks. *Adv. Mater. Technol.*, 1(4):1600047, July 2016.

[456] Pengli Yan, Emery Brown, Qing Su, Jun Li, Jian Wang, Changxue Xu, Chi Zhou, and Dong Lin. 3D printing hierarchical silver nanowire aerogel with highly compressive resilience and tensile elongation through tunable Poisson's ratio. *Small*, 13(38):1701756, Oct 2017.

[457] Guanglei Zhao, Chi Zhou, and Dong Lin. Tool path planning for directional freezing-based three-dimensional printing of nanomaterials. *J. Micro Nano-Manuf.*, 6(1):010905, Mar 2018.

[458] Taeil Kim, Chao Bao, Michael Hausmann, Gilberto Siqueira, Tanja Zimmermann, and Woo Soo Kim. 3D printed disposable wireless ion sensors with biocompatible cellulose composites. *Adv. Electron. Mater.*, 5(2):1800778, Feb 2019.

[459] Mark A. Skylar-Scott, Suman Gunasekaran, and Jennifer A. Lewis. Laser-assisted direct ink writing of planar and 3D metal architectures. *Proc. Natl. Acad. Sci.*, 113(22):6137–6142, May 2016.

[460] Ryan B. Wicker and Eric W. MacDonald. Multi-material, multi-technology stereolithography: this feature article covers a decade of research into tackling one of the major challenges of the stereolithography technique, which is including multiple materials in one construct. *Virtual Phys. Prototyp.*, 7(3):181–194, Sept 2012.

[461] Voltera Inc. Voltera V-One. www.voltera.io.

[462] Qijin Huang, Wenfeng Shen, Qingsong Xu, Ruiqin Tan, and Weijie Song. Properties of polyacrylic acid-coated silver nanoparticle ink for inkjet printing conductive tracks on paper with high conductivity. *Mater. Chem. Phys.*, 147(3):550–556, Oct 2014.

[463] Jung-Tang Wu, Steve Lien-Chung Hsu, Ming-Hsiu Tsai, Yu-Feng Liu, and Weng-Sing Hwang. Direct ink-jet printing of silver nitrate–silver nanowire hybrid inks to fabricate silver conductive lines. *J. Mater. Chem.*, 22(31):15599, 2012.

[464] Aurore Denneulin, Julien Bras, Fiona Carcone, Charles Neuman, and Anne Blayo. Impact of ink formulation on carbon nanotube network organization within inkjet printed conductive films. *Carbon*, 49(8):2603–2614, July 2011.

[465] Guo Liang Goh, Shweta Agarwala, and Wai Yee Yeong. Aerosol-jet-printed preferentially aligned carbon nanotube twin-lines for printed electronics. *ACS Appl. Mater. Interfaces*, 11(46):43719–43730, Nov 2019.

[466] Twinkle Pandhi, Eric Kreit, Roberto Aga, Kiyo Fujimoto, Mohammad Taghi Sharbati, Samane Khademi, A. Nicole Chang, Feng Xiong, Jessica Koehne, Emily M. Heckman, and David Estrada. Electrical transport and power dissipation in aerosol-jet-printed graphene interconnects. *Sci. Rep.*, 8(1):10842, Dec 2018.

[467] Garret McKerricher, Don Titterington, and Atif Shamim. A fully inkjet-printed 3-D honeycomb-inspired patch antenna. *IEEE Antennas Wirel. Propag. Lett.*, 15:544–547, 2016.

[468] Riikka Mikkonen, Paula Puistola, Ilari Jönkkäri, and Matti Mäntysalo. Inkjet printable polydimethylsiloxane for all-inkjet-printed multilayered soft electrical applications. *ACS Appl. Mater. Interfaces*, 12(10):11990–11997, Mar 2020.

[469] Nano Dimension. Nano Dimension DragonFly 3D printer. www.nano-di.com.

[470] Optomec, Inc. Optomec aerosol jet printing systems. optomec.com.

[471] Nazli Turan, Mortaza Saeidi-Javash, Jiahao Chen, Minxiang Zeng, Yanliang Zhang, and David B. Go. Atmospheric pressure and ambient temperature plasma jet sintering of aerosol jet printed silver nanoparticles. *ACS Appl. Mater. Interfaces*, 13(39):47244–47251, Oct 2021.

[472] Taibur Rahman, Luke Renaud, Deuk Heo, Michael Renn, and Rahul Panat. Aerosol based direct-write micro-additive fabrication method for sub-mm 3D metal-dielectric structures. *J. Micromech. Microeng.*, 25(10):107002, Oct 2015.

[473] Byeong Wan An, Kukjoo Kim, Heejoo Lee, So-Yun Kim, Yulhui Shim, Dae-Young Lee, Jun Yeob Song, and Jang-Ung Park. High-resolution printing of 3D structures using an electrohydrodynamic inkjet with multiple functional inks. *Adv. Mater.*, 27(29):4322–4328, Aug 2015.

[474] Won-Geun Kim, Jong-Min Lee, Younghwan Yang, Hongyoon Kim, Vasanthan Devaraj, Minjun Kim, Hyuk Jeong, Eun-Jung Choi, Jihyuk Yang, Yudong Jang, Trevon Badloe, Donghan Lee, Junsuk Rho, Ji Tae Kim, and Jin-Woo Oh. Three-dimensional plasmonic nanocluster-driven light–matter interaction for photoluminescence enhancement and picomolar-level biosensing. *Nano Lett.*, 22(12):4702–4711, June 2022.

[475] Seung Hwan Ko, Jaewon Chung, Nico Hotz, Koo Hyun Nam, and Costas P. Grigoropoulos. Metal nanoparticle direct inkjet printing for low-temperature 3D micro metal structure fabrication. *J. Micromech. Microeng.*, 20(12):125010, Dec 2010.

[476] Carmen Kullmann, Niklas C. Schirmer, Ming-Tsang Lee, Seung Hwan Ko, Nico Hotz, Costas P. Grigoropoulos, and Dimos Poulikakos. 3D micro-structures by piezoelectric inkjet printing of gold nanofluids. *J. Micromech. Microeng.*, 22(5):055022, May 2012.

[477] Jacob A. Sadie and Vivek Subramanian. Three-dimensional inkjet-printed interconnects using functional metallic nanoparticle inks. *Adv. Funct. Mater.*, 24(43):6834–6842, Nov 2014.

[478] Miriam Seiti, Olivier Degryse, Rosalba Monica Ferraro, Silvia Giliani, Veerle Bloemen, and Eleonora Ferraris. 3D Aerosol Jet® printing for microstructuring: advantages and limitations. *Int. J. Bioprinting*, 9(6):0257, June 2023.

[479] Mohammad Sadeq Saleh, Chunshan Hu, and Rahul Panat. Three-dimensional microarchitected materials and devices using nanoparticle assembly by pointwise spatial printing. *Sci. Adv.*, 3(3):e1601986, Mar 2017.

[480] Mohammad Sadeq Saleh, Jie Li, Jonghyun Park, and Rahul Panat. 3D printed hierarchically-porous microlattice electrode materials for exceptionally high specific capacity and areal capacity lithium ion batteries. *Addit. Manuf.*, 23:70–78, Oct 2018.

[481] Mohammad Sadeq Saleh, Chunshan Hu, Jacob Brenneman, Al Muntasar Al Mutairi, and Rahul Panat. 3D printed three-dimensional metallic microlattices with controlled and tunable mechanical properties. *Addit. Manuf.*, 39:101856, Mar 2021.

[482] Jun Ho Yu, Yoonsoo Rho, Heuiseok Kang, Hyun Suk Jung, and Kyung-Tae Kang. Electrical behavior of laser-sintered Cu based metal-organic decomposition ink in air environment and application as current collectors in supercapacitor. *Int. J. Precis. Eng. Manuf.-Green Technol.*, 2(4):333–337, Oct 2015.

[483] Yousef Farraj, Ariel Smooha, Alexander Kamyshny, and Shlomo Magdassi. Plasma-induced decomposition of copper complex ink for the formation of highly conductive copper tracks on heat-sensitive substrates. *ACS Appl. Mater. Interfaces*, 9(10):8766–8773, Mar 2017.

[484] Mohammad Vaseem, Garret McKerricher, and Atif Shamim. Robust design of a particle-free silver-organo-complex ink with high conductivity and inkjet stability for flexible electronics. *ACS Appl. Mater. Interfaces*, 8(1):177–186, Jan 2016.

[485] Arnold J. Kell, Chantal Paquet, Olga Mozenson, Iden Djavani-Tabrizi, Bhavana Deore, Xiangyang Liu, Gregory P. Lopinski, Robert James, Khelifa Hettak, Jafar Shaker, Adrian Momciu, Julie Ferrigno, Olivier Ferrand, Jian Xiong Hu, Sylvie Lafrenière, and Patrick R. L. Malenfant. Versatile molecular silver ink platform for printed flexible electronics. *ACS Appl. Mater. Interfaces*, 9(20):17226–17237, May 2017.

[486] Bing Zhang, Yong-Xin Guo, Hucheng Sun, and Yanjie Wu. Metallic, 3D-printed, k-band-stepped, double-ridged square horn antennas. *Appl. Sci.*, 8(1):33, Dec 2017.

[487] Guan-Long Huang, Shi-Gang Zhou, Tan-Huat Chio, and Tat-Soon Yeo. Fabrication of a high-efficiency waveguide antenna array via direct metal laser sintering. *IEEE Antennas Wirel. Propag. Lett.*, 15:622–625, 2016.

[488] Guan-Long Huang, Shi-Gang Zhou, and Tan-Huat Chio. Lightweight perforated horn antenna enabled by 3-d metal-direct-printing. In *2016 IEEE international symposium on antennas and propagation (APSURSI)*, pp. 481–482, Fajardo, PR, USA, June 2016. IEEE.

[489] Zhichao Li, Zhanhua Wang, Xinpeng Gan, Daihua Fu, Guoxia Fei, and Hesheng Xia. Selective laser sintering 3D printing: a way to construct 3D electrically conductive segregated network in polymer matrix. *Macromol. Mater. Eng.*, 302(11):1700211, Nov 2017.

[490] Xinpeng Gan, Jinzhi Wang, Zhanhua Wang, Zhuo Zheng, Marino Lavorgna, Alfredo Ronca, Guoxia Fei, and Hesheng Xia. Simultaneous realization of conductive segregation network microstructure and minimal surface porous macrostructure by SLS 3D printing. *Mater. Des.*, 178:107874, Sept 2019.

[491] Tim Hupfeld, Soma Salamon, Joachim Landers, Alexander Sommereyns, Carlos Doñate-Buendía, Jochen Schmidt, Heiko Wende, Michael Schmidt, Stephan Barcikowski, and Bilal Gökce. 3D printing of magnetic parts by laser powder bed fusion of iron oxide nanoparticle functionalized polyamide powders. *J. Mater. Chem. C*, 8(35):12204–12217, 2020.

[492] Hongzhi Wu, Ouyangxu Wang, Yujia Tian, Mingzhe Wang, Bin Su, Chunze Yan, Kun Zhou, and Yusheng Shi. Selective laser sintering-based 4D printing of magnetism-responsive grippers. *ACS Appl. Mater. Interfaces*, 13(11):12679–12688, Mar 2021.

[493] Seung Hwan Ko, Jaewon Chung, Heng Pan, Costas P. Grigoropoulos, and Dimos Poulikakos. Fabrication of multilayer passive and active electric components on polymer using inkjet printing and low temperature laser processing. *Sens. Actuators Phys.*, 134(1):161–168, Feb 2007.

[494] Patrick F. Flowers, Christopher Reyes, Shengrong Ye, Myung Jun Kim, and Benjamin J. Wiley. 3D printing electronic components and circuits with conductive thermoplastic filament. *Addit. Manuf.*, 18:156–163, Dec 2017.

[495] Garret McKerricher, Jose Gonzalez Perez, and Atif Shamim. Fully inkjet printed RF inductors and capacitors using polymer dielectric and silver conductive ink with through vias. *IEEE Trans. Electron Devices*, 62(3):1002–1009, Mar 2015.

[496] Shuang-zhuang Guo, Xuelu Yang, Marie-Claude Heuzey, and Daniel Therriault. 3D printing of a multifunctional nanocomposite helical liquid sensor. *Nanoscale*, 7(15):6451–6456, 2015.

[497] Devin J. Roach, Christopher Roberts, Janet Wong, Xiao Kuang, Joshua Kovitz, Qiang Zhang, Thomas G. Spence, and H. Jerry Qi. Surface modification of fused filament fabrication (FFF) 3D printed substrates by inkjet printing polyimide for printed electronics. *Addit. Manuf.*, 36:101544, Dec 2020.

[498] K. Prashantha and F. Roger. Multifunctional properties of 3D printed poly(lactic acid)/graphene nanocomposites by fused deposition modeling. *J. Macromol. Sci. Part A*, 54(1):24–29, Jan 2017.

[499] Jinpil Tak, Do-Gu Kang, and Jaehoon Choi. A lightweight waveguide horn antenna made via 3 D printing and conductive spray coating. *Microw. Opt. Technol. Lett.*, 59(3):727–729, Mar 2017.

[500] M. Kong, G. Shin, S.-H. Lee, and I.-J Yoon. Electrically small folded spherical helix antennas using copper strips and 3D printing technology. *Electron. Lett.*, 52(12):994–996, June 2016.

[501] Nanjia Zhou, Chengye Liu, Jennifer A. Lewis, and Donhee Ham. Gigahertz electromagnetic structures via direct ink writing for radio-frequency oscillator and transmitter applications. *Adv. Mater.*, 29(15):1605198, Apr 2017.

[502] Jacob J. Adams, Eric B. Duoss, Thomas F. Malkowski, Michael J. Motala, Bok Yeop Ahn, Ralph G. Nuzzo, Jennifer T. Bernhard, and Jennifer A. Lewis. Conformal printing of electrically small antennas on three-dimensional surfaces. *Adv. Mater.*, 23(11):1335–1340, Mar 2011.

[503] Wooik Jung, Yoon-Ho Jung, Peter V. Pikhitsa, Jicheng Feng, Younghwan Yang, Minkyung Kim, Hao-Yuan Tsai, Takuo Tanaka, Jooyeon Shin, Kwang-Yeong Kim, Hoseop Choi, Junsuk Rho, and Mansoo Choi. Three-dimensional nanoprinting via charged aerosol jets. *Nature*, 592(7852):54–59, Apr 2021.

[504] Yaru Wang, Xueqin Zhang, Ruyue Su, Mingji Chen, Chujing Shen, Hao Xu, and Rujie He. 3D printed antennas for 5G communication: current progress and future challenges. *Chin. J. Mech. Eng. Addit. Manuf. Front.*, 2(1):100065, Mar 2023.

[505] Pochiang Chen, Yue Fu, Radnoosh Aminirad, Chuan Wang, Jialu Zhang, Kang Wang, Kosmas Galatsis, and Chongwu Zhou. Fully printed separated carbon nanotube thin film transistor circuits and its application in organic light emitting diode control. *Nano Lett.*, 11(12):5301–5308, Dec 2011.

[506] Suoming Zhang, Le Cai, Tongyu Wang, Jinshui Miao, Nelson Sepúlveda, and Chuan Wang. Fully printed flexible carbon nanotube photodetectors. *Appl. Phys. Lett.*, 110(12):123105, Mar 2017.

[507] Der-Hsien Lien, Zhen-Kai Kao, Teng-Han Huang, Ying-Chih Liao, Si-Chen Lee, and Jr-Hau He. All-printed paper memory. *ACS Nano*, 8(8):7613–7619, Aug 2014.

[508] Jonathan Harrop. *3D printing 2016–2026: technologies, markets, players*. July 2016. https://www.idtechex.com/en/research-report/3d-printing-2016-2026-technologies-markets-players/485.

[509] Jiaxin Fan, Carlo Montemagno, and Manisha Gupta. 3D printed high transconductance organic electrochemical transistors on flexible substrates. *Org. Electron.*, 73:122–129, Oct 2019.

[510] Tanyaradzwa N. Mangoma, Shunsuke Yamamoto, George G. Malliaras, and Ronan Daly. Hybrid 3D/inkjet-printed organic neuromorphic transistors. *Adv. Mater. Technol.*, 7(2):2000798, Feb 2022.

[511] Valentina Bertana, Giorgio Scordo, Matteo Parmeggiani, Luciano Scaltrito, Sergio Ferrero, Manuel Gomez Gomez, Matteo Cocuzza, Davide Vurro, Pasquale D'Angelo, Salvatore Iannotta, Candido F. Pirri, and Simone L. Marasso. Rapid prototyping of 3D organic electrochemical transistors by composite photocurable resin. *Sci. Rep.*, 10(1):13335, Aug 2020.

[512] Xingang Liu, Yinghao Shang, Jihai Zhang, and Chuhong Zhang. Ionic liquid-assisted 3D printing of self-polarized β-PVDF for flexible piezoelectric energy harvesting. *ACS Appl. Mater. Interfaces*, 13(12):14334–14341, Mar 2021.

[513] Hei Wong and Kuniyuki Kakushima. On the vertically stacked gate-all-around nanosheet and nanowire transistor scaling beyond the 5 nm technology node. *Nanomaterials*, 12(10):1739, May 2022.

[514] Raj Kumar, Shashi Bala, and Arvind Kumar. Study and analysis of advanced 3D multi-gate junctionless transistors. *Silicon*, 14(3):1053–1067, Feb 2022.

[515] S. Monfray, T. Skotnicki, Y. Morand, S. Descombes, P. Coronel, P. Mazoyer, S. Harrison, P. Ribot, A. Talbot, D. Dutartre, M. Haond, R. Palla, Y. Le Friec, F. Leverd, M.-E. Nier, C. Vizioz, and D. Louis. 50 nm-gate all around (GAA)-silicon on nothing (son)-devices: a simple way to co-integration of GAA transistors within bulk MOSFET process. In *2002 symposium on VLSI technology. Digest of technical papers (cat. No. 01CH37303)*, pp. 108–109, Honolulu, HI, USA, 2002. IEEE.

[516] Leona V. Lingstedt, Matteo Ghittorelli, Hao Lu, Dimitrios A. Koutsouras, Tomasz Marszalek, Fabrizio Torricelli, N. Irina Crăciun, Paschalis Gkoupidenis, and Paul W. M. Blom. Effect of DMSO solvent treatments on the performance of PEDOT:PSS based organic electrochemical transistors. *Adv. Electron. Mater.*, 1800804, Feb 2019.

[517] Yu Kimura, Takashi Nagase, Takashi Kobayashi, Azusa Hamaguchi, Yoshinori Ikeda, Takashi Shiro, Kazuo Takimiya, and Hiroyoshi Naito. Soluble organic semiconductor precursor with specific phase separation for high-performance printed organic transistors. *Adv. Mater.*, 27(4):727–732, Jan 2015.

[518] Muhammad R. Niazi, Ruipeng Li, Er Qiang Li, Ahmad R. Kirmani, Maged Abdelsamie, Qingxiao Wang, Wenyang Pan, Marcia M. Payne, John E. Anthony, Detlef-M. Smilgies, Sigurdur T. Thoroddsen, Emmanuel P. Giannelis, and Aram Amassian. Solution-printed organic semiconductor blends exhibiting transport properties on par with single crystals. *Nat. Commun.*, 6(1):8598, Nov 2015.

[519] William J. Scheideler, Rajan Kumar, Andre R. Zeumault, and Vivek Subramanian. Low-temperature-processed printed metal oxide transistors based on pure aqueous inks. *Adv. Funct. Mater.*, 27(14):1606062, Apr 2017.

[520] S.-Y. Kim, K. Kim, Y. H. Hwang, J. Park, J. Jang, Y. Nam, Y. Kang, M. Kim, H. J. Park, Z. Lee, J. Choi, Y. Kim, S. Jeong, B.-S. Bae, and J.-U. Park. High-resolution electrohydrodynamic inkjet printing of stretchable metal oxide semiconductor transistors with high performance. *Nanoscale*, 8(39):17113–17121, 2016.

[521] Zhengshang Wang, Wen Cui, Hao Yuan, Xiaoli Kang, Zhou Zheng, Wenbin Qiu, Qiujun Hu, Jun Tang, and Xudong Cui. Direct ink writing of Bi2Te3-based thermoelectric materials induced by rheological design. *Mater. Today Energy*, 31:101206, Jan 2023.

[522] Fredrick Kim, Seong Eun Yang, Hyejin Ju, Seungjun Choo, Jungsoo Lee, Gyeonghun Kim, Soo-ho Jung, Suntae Kim, Chaenyung Cha, Kyung Tae Kim, Sangjoon Ahn, Han Gi Chae, and Jae Sung Son. Direct ink writing of three-dimensional thermoelectric microarchitectures. *Nat. Electron.*, 4(8):579–587, Aug 2021.

[523] Byeong Wan An, Kukjoo Kim, Mijung Kim, So-Yun Kim, Seung-Hyun Hur, and Jang-Ung Park. Direct printing of reduced graphene oxide on planar or highly curved surfaces with high resolutions using electrohydrodynamics. *Small*, 11(19):2263–2268, May 2015.

[524] Peter Mack Grubb, Harish Subbaraman, Saungeun Park, Deji Akinwande, and Ray T. Chen. Inkjet printing of high performance transistors with micron order chemically set gaps. *Sci. Rep.*, 7(1):1202, Apr 2017.

[525] Junfeng Sun, Ashish Sapkota, Hyejin Park, Prince Wesley, Younsu Jung, Bijendra Bishow Maskey, Yushin Kim, Yutaka Majima, Jianfu Ding, Jianying Ouyang, Chang Guo, Jacques Lefebvre, Zhao Li, Patrick R. L. Malenfant, Ali Javey, and Gyoujin Cho. Fully R2R-printed carbon-nanotube-based limitless length of flexible active-matrix for electrophoretic display application. *Adv. Electron. Mater.*, 6(4):1901431, Apr 2020.

[526] Carissa S. Jones, Xuejun Lu, Mike Renn, Mike Stroder, and Wu-Sheng Shih. Aerosol-jet-printed, high-speed, flexible thin-film transistor made using single-walled carbon nanotube solution. *Microelectron. Eng.*, 87(3):434–437, Mar 2010.

[527] Zhenwei Wang, Hyunho Kim, and Husam N. Alshareef. Oxide thin-film electronics using all-MXene electrical contacts. *Adv. Mater.*, 30(15):1706656, Apr 2018.

[528] Le Cai, Suoming Zhang, Jinshui Miao, Zhibin Yu, and Chuan Wang. Fully printed stretchable thin-film transistors and integrated logic circuits. *ACS Nano*, 10(12):11459–11468, Dec 2016.

[529] Kenjiro Fukuda, Yasunori Takeda, Yudai Yoshimura, Rei Shiwaku, Lam Truc Tran, Tomohito Sekine, Makoto Mizukami, Daisuke Kumaki, and Shizuo Tokito. Fully-printed high-performance organic thin-film transistors and circuitry on one-micron-thick polymer films. *Nat. Commun.*, 5(1):4147, June 2014.

[530] John Biggs, James Myers, Jedrzej Kufel, Emre Ozer, Simon Craske, Antony Sou, Catherine Ramsdale, Ken Williamson, Richard Price, and Scott White. A natively flexible 32-bit Arm microprocessor. *Nature*, 595(7868):532–536, July 2021.

[531] Xiaojin Peng, Jian Yuan, Shirley Shen, Mei Gao, Anthony S. R. Chesman, Hong Yin, Jinshu Cheng, Qi Zhang, and Dechan Angmo. Perovskite and organic solar cells fabricated by inkjet printing: progress and prospects. *Adv. Funct. Mater.*, 27(41):1703704, Nov 2017.

[532] Guodong Wang, Muhammad Abdullah Adil, Jianqi Zhang, and Zhixiang Wei. Large-area organic solar cells: material requirements, modular designs, and printing methods. *Adv. Mater.*, 31(45):1805089, Nov 2019.

[533] Xin-Yi Zeng, Yan-Qing Tang, Xiao-Yi Cai, Jian-Xin Tang, and Yan-Qing Li. Solution-processed oleds for printing displays. *Mater. Chem. Front.*, 7(7):1166–1196, 2023.

[534] Zhongyuan Wu, Liangchen Yan, Yongqian Li, Xuehuan Feng, Huaiting Shih, Taejin Kim, Yuqing Peng, Jianwei Yu, and Xue Dong. Development of 55-in. 8k amoled tv based on coplanar oxide thin-film transistors and inkjet printing process. *J. Soc. Inf. Disp.*, 28(5):418–427, May 2020.

[535] Johannes Zimmermann, Stefan Schlisske, Martin Held, Jean-Nicolas Tisserant, Luca Porcarelli, Ana Sanchez-Sanchez, David Mecerreyes, and Gerardo Hernandez-Sosa. Ultrathin fully printed light-emitting electrochemical cells with arbitrary designs on biocompatible substrates. *Adv. Mater. Technol.*, 4(3):1800641, Mar 2019.

[536] Xueying Xiong, Changting Wei, Liming Xie, Ming Chen, Pengyu Tang, Wei Shen, Zhengtao Deng, Xia Li, Yongjie Duan, Wenming Su, Haibo Zeng, and Zheng Cui. Realizing 17.0 % external quantum efficiency in red quantum dot light-emitting diodes by pursuing the ideal inkjet-printed film and interface. *Org. Electron.*, 73:247–254, Oct 2019.

[537] Giovanni Azzellino, Francesca S. Freyria, Michel Nasilowski, Moungi G. Bawendi, and Vladimir Bulović. Micron-scale patterning of high quantum yield quantum dot leds. *Adv. Mater. Technol.*, 4(7):1800727, July 2019.

[538] Menghua Zhu, Yongqing Duan, Nian Liu, Hegeng Li, Jinghui Li, Peipei Du, Zhifang Tan, Guangda Niu, Liang Gao, YongAn Huang, Zhouping Yin, and Jiang Tang. Electrohydrodynamically printed high-resolution full-color hybrid perovskites. *Adv. Funct. Mater.*, 29(35):1903294, Aug 2019.

[539] Congbiao Jiang, Zhiming Zhong, Baiquan Liu, Zhiwei He, Jianhua Zou, Lei Wang, Jian Wang, JunBiao Peng, and Yong Cao. Coffee-ring-free quantum dot thin film using inkjet printing from a mixed-solvent system on modified ZnO transport layer for light-emitting devices. *ACS Appl. Mater. Interfaces*, 8(39):26162–26168, Oct 2016.

[540] Lifu Shi, Linghai Meng, Feng Jiang, Yong Ge, Fei Li, Xian-gang Wu, and Haizheng Zhong. In situ inkjet printing strategy for fabricating perovskite quantum dot patterns. *Adv. Funct. Mater.*, 29(37):1903648, Sept 2019.

[541] Sjoerd A. Veldhuis, Pablo P. Boix, Natalia Yantara, Mingjie Li, Tze Chien Sum, Nripan Mathews, and Subodh G. Mhaisalkar. Perovskite materials for light-emitting diodes and lasers. *Adv. Mater.*, 28(32):6804–6834, Aug 2016.

[542] Gabriel Loke, Rodger Yuan, Michael Rein, Tural Khudiyev, Yash Jain, John Joannopoulos, and Yoel Fink. Structured multimaterial filaments for 3D printing of optoelectronics. *Nat. Commun.*, 10(1):4010, Sept 2019.

[543] Jongcheon Bae, Sanghyeon Lee, Jinhyuck Ahn, Jung Hyun Kim, Muhammad Wajahat, Won Suk Chang, Seog-Young Yoon, Ji Tae Kim, Seung Kwon Seol, and Jaeyeon Pyo. 3D-printed quantum dot nanopixels. *ACS Nano*, 14(9):10993–11001, Sept 2020.

[544] Ian A. Howard, Tobias Abzieher, Ihteaz M. Hossain, Helge Eggers, Fabian Schackmar, Simon Ternes, Bryce S. Richards, Uli Lemmer, and Ulrich W. Paetzold. Coated and printed perovskites for photovoltaic applications. *Adv. Mater.*, 31(26):1806702, June 2019.

[545] Aron J. Huckaba, Yonghui Lee, Rui Xia, Sanghyun Paek, Victor Costa Bassetto, Emad Oveisi, Andreas Lesch, Sachin Kinge, Paul J. Dyson, Hubert Girault, and Mohammad Khaja Nazeeruddin. Inkjet-printed mesoporous TiO 2 and perovskite layers for high efficiency perovskite solar cells. *Energy Technol.*, 7(2):317–324, Feb 2019.

[546] Florian Mathies, Helge Eggers, Bryce S. Richards, Gerardo Hernandez-Sosa, Uli Lemmer, and Ulrich W. Paetzold. Inkjet-printed triple cation perovskite solar cells. *ACS Appl. Energy Mater.*, 1(5):1834–1839, May 2018.

[547] Anyi Mei, Xiong Li, Linfeng Liu, Zhiliang Ku, Tongfa Liu, Yaoguang Rong, Mi Xu, Min Hu, Jiangzhao Chen, Ying Yang, Michael Grätzel, and Hongwei Han. A hole-conductor-free, fully printable mesoscopic perovskite solar cell with high stability. *Science*, 345(6194):295–298, July 2014.

[548] Yaoguang Rong, Yue Hu, Anyi Mei, Hairen Tan, Makhsud I. Saidaminov, Sang Il Seok, Michael D. McGehee, Edward H. Sargent, and Hongwei Han. Challenges for commercializing perovskite solar cells. *Science*, 361(6408):eaat8235, Sept 2018.

[549] Dian Wang, Matthew Wright, Naveen Kumar Elumalai, and Ashraf Uddin. Stability of perovskite solar cells. *Sol. Energy Mater. Sol. Cells*, 147:255–275, Apr 2016.

[550] Anastasiia Glushkova, Pavao Andričević, Rita Smajda, Bálint Náfrádi, Márton Kollár, Veljko Djokić, Alla Arakcheeva, László Forró, Raphael Pugin, and Endre Horváth. Ultrasensitive 3D aerosol-jet-printed perovskite x-ray photodetector. *ACS Nano*, 15(3):4077–4084, Mar 2021.

[551] Chunhe Yang, Erjun Zhou, Shoji Miyanishi, Kazuhito Hashimoto, and Keisuke Tajima. Preparation of active layers in polymer solar cells by aerosol jet printing. *ACS Appl. Mater. Interfaces*, 3(10):4053–4058, Oct 2011.

[552] Anubha A. Gupta, Shivaram Arunachalam, Sylvain G. Cloutier, and Ricardo Izquierdo. Fully aerosol-jet printed, high-performance nanoporous ZnO ultraviolet photodetectors. *ACS Photonics*, 5(10):3923–3929, Oct 2018.

[553] Pälvi Kopola, Birger Zimmermann, Aleksander Filipovic, Hans-Frieder Schleiermacher, Johannes Greulich, Sanna Rousu, Jukka Hast, Risto Myllylä, and Uli Würfel. Aerosol jet printed grid for ITO-free inverted organic solar cells. *Sol. Energy Mater. Sol. Cells*, 107:252–258, Dec 2012.

[554] A. Mette, P. L. Richter, M. Hörteis, and S. W. Glunz. Metal aerosol jet printing for solar cell metallization. *Prog. Photovolt. Res. Appl.*, 15(7):621–627, Nov 2007.

[555] G. A. Dosovitskiy, P. V. Karpyuk, P. V. Evdokimov, D. E. Kuznetsova, V. A. Mechinsky, A. E. Borisevich, A. A. Fedorov, V. I. Putlayev, A. E. Dosovitskiy, and M. V. Korjik. First 3D-printed complex inorganic polycrystalline scintillator. *CrystEngComm*, 19(30):4260–4264, 2017.

[556] Ł. Kapłon, D. Kulig, S. Beddar, T. Fiutowski, W. Górska, J. Hajduga, P. Jurgielewicz, D. Kabat, K. Kalecińska, M. Kopeć, S. Koperny, B. Mindur, J. Moroń, G. Moskal, S. Niedźwiecki, M. Silarski, F. Sobczuk, T. Szumlak, and A. Ruciński. Investigation of the light output of 3D-printed plastic scintillators for dosimetry applications. *Radiat. Meas.*, 158:106864, Nov 2022.

[557] Bin Bao, Mingzhu Li, Yuan Li, Jieke Jiang, Zhenkun Gu, Xingye Zhang, Lei Jiang, and Yanlin Song. Patterning fluorescent quantum dot nanocomposites by reactive inkjet printing. *Small*, 11(14):1649–1654, Apr 2015.

[558] Mojun Chen, Jihyuk Yang, Zhenyu Wang, Zhaoyi Xu, Heekwon Lee, Hyeonseok Lee, Zhiwen Zhou, Shien-Ping Feng, Sanghyeon Lee, Jaeyeon Pyo, Seung Kwon Seol, Dong-Keun Ki, and Ji Tae Kim. 3D nanoprinting of perovskites. *Adv. Mater.*, 31(44):1904073, Nov 2019.

[559] Qingchuan Guo, Reza Ghadiri, Thomas Weigel, Andreas Aumann, Evgeny Gurevich, Cemal Esen, Olaf Medenbach, Wei Cheng, Boris Chichkov, and Andreas Ostendorf. Comparison of in situ and ex situ methods for synthesis of two-photon polymerization polymer nanocomposites. *Polymers*, 6(7):2037–2050, July 2014.

[560] Andrey Vyatskikh, Stéphane Delalande, Akira Kudo, Xuan Zhang, Carlos M. Portela, and Julia R. Greer. Additive manufacturing of 3D nano-architected metals. *Nat. Commun.*, 9(1):593, Dec 2018.

[561] Redouane Krini, Cheol Woo Ha, Prem Prabhakaran, Hicham El Mard, Dong-Yol Yang, Rudolf Zentel, and Kwang-Sup Lee. Photosensitive functionalized surface-modified quantum dots for polymeric structures via two-photon-initiated polymerization technique. *Macromol. Rapid Commun.*, 36(11):1108–1114, June 2015.

[562] Bijal B. Patel, Dylan J. Walsh, Do Hoon Kim, Justin Kwok, Byeongdu Lee, Damien Guironnet, and Ying Diao. Tunable structural color of bottlebrush block copolymers through direct-write 3D printing from solution. *Sci. Adv.*, 6(24):eaaz7202, June 2020.

[563] Nanjia Zhou, Yehonadav Bekenstein, Carissa N. Eisler, Dandan Zhang, Adam M. Schwartzberg, Peidong Yang, A. Paul Alivisatos, and Jennifer A. Lewis. Perovskite nanowire–block copolymer composites with digitally programmable polarization anisotropy. *Sci. Adv.*, 5(5):eaav8141, May 2019.

[564] V. Harinarayana and Y. C. Shin. Two-photon lithography for three-dimensional fabrication in micro/nanoscale regime: a comprehensive review. *Opt. Laser Technol.*, 142:107180, Oct 2021.

[565] Callum Vidler, Kenneth Crozier, and David Collins. Ultra-resolution scalable microprinting. *Microsyst. Nanoeng.*, 9(1):67, s41378-023-00537-9, May 2023.

[566] Zhenyu Wang, Xu Liu, Xi Shen, Ne Myo Han, Ying Wu, Qingbin Zheng, Jingjing Jia, Ning Wang, and Jang-Kyo Kim. An ultralight graphene honeycomb sandwich for stretchable light-emitting displays. *Adv. Funct. Mater.*, 28(19):1707043, May 2018.

[567] Xin Zhao, Beatriz Mendoza Sánchez, Peter J. Dobson, and Patrick S. Grant. The role of nanomaterials in redox-based supercapacitors for next generation energy storage devices. *Nanoscale*, 3(3):839, 2011.

[568] Martin Pumera. Graphene-based nanomaterials for energy storage. *Energy Environ. Sci.*, 4(3):668–674, 2011.

[569] Yu-Guo Guo, Jin-Song Hu, and Li-Jun Wan. Nanostructured materials for electrochemical energy conversion and storage devices. *Adv. Mater.*, 20(15):2878–2887, Aug 2008.

[570] Seong Jin An, Jianlin Li, Claus Daniel, Debasish Mohanty, Shrikant Nagpure, and David L. Wood. The state of understanding of the lithium-ion-battery graphite solid electrolyte interphase (SEI) and its relationship to formation cycling. *Carbon*, 105:52–76, Aug 2016.

[571] Madhav Singh, Jörg Kaiser, and Horst Hahn. Thick electrodes for high energy lithium ion batteries. *J. Electrochem. Soc.*, 162(7):A1196–A1201, 2015.

[572] Adilet Zhakeyev, Panfeng Wang, Li Zhang, Wenmiao Shu, Huizhi Wang, and Jin Xuan. Additive manufacturing: unlocking the evolution of energy materials. *Adv. Sci.*, 4(10):1700187, Oct 2017.

[573] Wei Yu, Han Zhou, Ben Q. Li, and Shujiang Ding. 3D printing of carbon nanotubes-based microsupercapacitors. *ACS Appl. Mater. Interfaces*, 9(5):4597–4604, Feb 2017.

[574] Bin Bian, Dai Shi, Xiaobing Cai, Mingjun Hu, Qiuquan Guo, Chuhong Zhang, Qi Wang, Andy Xueliang Sun, and Jun Yang. 3D printed porous carbon anode for enhanced power generation in microbial fuel cell. *Nano Energy*, 44:174–180, Feb 2018.

[575] Fredrick Kim, Beomjin Kwon, Youngho Eom, Ji Eun Lee, Sangmin Park, Seungki Jo, Sung Hoon Park, Bong-Seo Kim, Hye Jin Im, Min Ho Lee, Tae Sik Min, Kyung Tae Kim, Han Gi Chae, William P. King, and Jae Sung Son. 3D printing of shape-conformable thermoelectric materials using all-inorganic Bi2Te3-based inks. *Nat. Energy*, 3(4):301–309, Apr 2018.

[576] Kacper Skarżyński and Marcin Słoma. Printed electronics in radiofrequency energy harvesters and wireless power transfer rectennas for IoT applications. *Adv. Electron. Mater.*, 2300238, July 2023.

[577] Jin Yang, Yanshuo Sun, Jianjun Zhang, Baodong Chen, and Zhong Lin Wang. 3D-printed bearing structural triboelectric nanogenerator for intelligent vehicle monitoring. *Cell Rep. Phys. Sci.*, 2(12):100666, Dec 2021.

[578] Kun Fu, Yibo Wang, Chaoyi Yan, Yonggang Yao, Yanan Chen, Jiaqi Dai, Steven Lacey, Yanbin Wang, Jiayu Wan, Tian Li, Zhengyang Wang, Yue Xu, and Liangbing Hu. Graphene oxide-based electrode inks for 3D-printed lithium-ion batteries. *Adv. Mater.*, 28(13):2587–2594, Apr 2016.

[579] Ke Sun, Teng-Sing Wei, Bok Yeop Ahn, Jung Yoon Seo, Shen J. Dillon, and Jennifer A. Lewis. 3D printing of interdigitated Li-ion microbattery architectures. *Adv. Mater.*, 25(33):4539–4543, Sept 2013.

[580] Yibo Wang, Chaoji Chen, Hua Xie, Tingting Gao, Yonggang Yao, Glenn Pastel, Xiaogang Han, Yiju Li, Jiupeng Zhao, Kun (Kelvin) Fu, and Liangbing Hu. 3D-printed all-fiber Li-ion battery toward wearable energy storage. *Adv. Funct. Mater.*, 27(43):1703140, Nov 2017.

[581] Daxian Cao, Yingjie Xing, Karnpiwat Tantratian, Xiao Wang, Yi Ma, Alolika Mukhopadhyay, Zheng Cheng, Qing Zhang, Yucong Jiao, Lei Chen, and Hongli Zhu. 3D printed high-performance lithium metal microbatteries enabled by nanocellulose. *Adv. Mater.*, 31(14):1807313, Apr 2019.

[582] Xuejie Gao, Qian Sun, Xiaofei Yang, Jianneng Liang, Alicia Koo, Weihan Li, Jianwen Liang, Jiwei Wang, Ruying Li, Frederick Benjamin Holness, Aaron David Price, Songlin Yang, Tsun-Kong Sham, and Xueliang Sun. Toward a remarkable Li-S battery via 3D printing. *Nano Energy*, 56:595–603, Feb 2019.

[583] Chanhoon Kim, Bok Yeop Ahn, Teng-Sing Wei, Yejin Jo, Sunho Jeong, Youngmin Choi, Il-Doo Kim, and Jennifer A. Lewis. High-power aqueous zinc-ion batteries for customized electronic devices. *ACS Nano*, 12(12):11838–11846, Dec 2018.

[584] Emery Brown, Pengli Yan, Halil Tekik, Ayyappan Elangovan, Jian Wang, Dong Lin, and Jun Li. 3D printing of hybrid MoS2-graphene aerogels as highly porous electrode materials for sodium ion battery anodes. *Mater. Des.*, 170:107689, May 2019.

[585] Kai Chi, Zheye Zhang, Jiangbo Xi, Yongan Huang, Fei Xiao, Shuai Wang, and Yunqi Liu. Freestanding graphene paper supported three-dimensional porous graphene–polyaniline nanocomposite synthesized by inkjet printing and in flexible all-solid-state supercapacitor. *ACS Appl. Mater. Interfaces*, 6(18):16312–16319, Sep 2014.

[586] Cheng Zhu, Tianyu Liu, Fang Qian, T. Yong-Jin Han, Eric B. Duoss, Joshua D. Kuntz, Christopher M. Spadaccini, Marcus A. Worsley, and Yat Li. Supercapacitors based on three-dimensional hierarchical graphene aerogels with periodic macropores. *Nano Lett.*, 16(6):3448–3456, June 2016.

[587] Victoria G. Rocha, Esther García-Tuñón, Cristina Botas, Foivos Markoulidis, Ezra Feilden, Eleonora D'Elia, Na Ni, Milo Shaffer, and Eduardo Saiz. Multimaterial 3D printing of graphene-based electrodes for electrochemical energy storage using thermoresponsive inks. *ACS Appl. Mater. Interfaces*, 9(42):37136–37145, Oct 2017.

[588] Wenji Yang, Jie Yang, Jae Jong Byun, Francis P. Moissinac, Jiaqi Xu, Sarah J. Haigh, Marco Domingos, Mark A. Bissett, Robert A. W. Dryfe, and Suelen Barg. 3D printing of freestanding MXene architectures for current-collector-free supercapacitors. *Adv. Mater.*, 31(37):1902725, Sept 2019.

[589] Lianghao Yu, Zhaodi Fan, Yuanlong Shao, Zhengnan Tian, Jingyu Sun, and Zhongfan Liu. Versatile n-doped MXene ink for printed electrochemical energy storage application. *Adv. Energy Mater.*, 9(34):1901839, Sept 2019.

[590] Feng Zhang, Tianyu Liu, Mingyang Li, Minghao Yu, Yang Luo, Yexiang Tong, and Yat Li. Multiscale pore network boosts capacitance of carbon electrodes for ultrafast charging. *Nano Lett.*, 17(5):3097–3104, May 2017.

[591] Steven D. Lacey, Dylan J. Kirsch, Yiju Li, Joseph T. Morgenstern, Brady C. Zarket, Yonggang Yao, Jiaqi Dai, Laurence Q. Garcia, Boyang Liu, Tingting Gao, Shaomao Xu, Srinivasa R. Raghavan, John W. Connell, Yi Lin, and Liangbing Hu. Extrusion-based 3D printing of hierarchically porous advanced battery electrodes. *Adv. Mater.*, 30(12):1705651, Mar 2018.

[592] Christopher W. Foster, Michael P. Down, Yan Zhang, Xiaobo Ji, Samuel J. Rowley-Neale, Graham C. Smith, Peter J. Kelly, and Craig E. Banks. 3D printed graphene based energy storage devices. *Sci. Rep.*, 7(1):42233, Sept 2017.

[593] Alexis Maurel, Matthieu Courty, Benoit Fleutot, Hugues Tortajada, Kalappa Prashantha, Michel Armand, Sylvie Grugeon, Stéphane Panier, and Loic Dupont. Highly loaded graphite–polylactic acid composite-based filaments for lithium-ion battery three-dimensional printing. *Chem. Mater.*, 30(21):7484–7493, Nov 2018.

[594] Chuan Yi Foo, Hong Ngee Lim, Mohd Adzir Mahdi, Mohd Haniff Wahid, and Nay Ming Huang. Three-dimensional printed electrode and its novel applications in electronic devices. *Sci. Rep.*, 8(1):7399, May 2018.

[595] Yang Yang, Zeyu Chen, Xuan Song, Benpeng Zhu, Tzung Hsiai, Pin-I Wu, Rui Xiong, Jing Shi, Yong Chen, Qifa Zhou, and K. Kirk Shung. Three dimensional printing of high dielectric capacitor using projection based stereolithography method. *Nano Energy*, 22:414–421, Apr 2016.

[596] Ethan B. Secor, Theodore Z. Gao, Ahmad E. Islam, Rahul Rao, Shay G. Wallace, Jian Zhu, Karl W. Putz, Benji Maruyama, and Mark C. Hersam. Enhanced conductivity, adhesion, and environmental stability of printed graphene inks with nitrocellulose. *Chem. Mater.*, 29(5):2332–2340, Mar 2017.

[597] Meng Cheng, Yizhou Jiang, Wentao Yao, Yifei Yuan, Ramasubramonian Deivanayagam, Tara Foroozan, Zhennan Huang, Boao Song, Ramin Rojaee, Tolou Shokuhfar, Yayue Pan, Jun Lu, and Reza Shahbazian-Yassar. Elevated-temperature 3D printing of hybrid solid-state electrolyte for Li-ion batteries. *Adv. Mater.*, 30(39):1800615, Sept 2018.

[598] Jiuk Jang, Hyobeom Kim, Sangyoon Ji, Ha Jun Kim, Min Soo Kang, Tae Soo Kim, Jong-eun Won, Jae-Hyun Lee, Jinwoo Cheon, Kibum Kang, Won Bin Im, and Jang-Ung Park. Mechanoluminescent, air-dielectric MoS 2 transistors as active-matrix pressure sensors for wide detection ranges from footsteps to cellular motions. *Nano Lett.*, 20(1):66–74, Jan 2020.

[599] Jayoung Kim, Juliane R. Sempionatto, Somayeh Imani, Martin C. Hartel, Abbas Barfidokht, Guangda Tang, Alan S. Campbell, Patrick P. Mercier, and Joseph Wang. Simultaneous monitoring of sweat and interstitial fluid using a single wearable biosensor platform. *Adv. Sci.*, 5(10):1800880, Oct 2018.

[600] Byungkook Oh, Young-Geun Park, Hwaebong Jung, Sangyoon Ji, Woon Hyung Cheong, Jinwoo Cheon, Wooyoung Lee, and Jang-Ung Park. Untethered soft robotics with fully integrated wireless sensing and actuating systems for somatosensory and respiratory functions. *Soft Robot.*, 7(5):564–573, Oct 2020.

[601] Ju Young Kim, Seulgi Ji, Sungmook Jung, Beyong-Hwan Ryu, Hyun-Suk Kim, Sun Sook Lee, Youngmin Choi, and Sunho Jeong. 3D printable composite dough for stretchable, ultrasensitive and body-patchable strain sensors. *Nanoscale*, 9(31):11035–11046, 2017.

[602] Yitian Wang, Qiang Chang, Rixing Zhan, Kaige Xu, Ying Wang, Xingying Zhang, Bingyun Li, Gaoxing Luo, Malcolm Xing, and Wen Zhong. Tough but self-healing and 3D printable hydrogels for e-skin, e-noses and laser controlled actuators. *J. Mater. Chem. A*, 7(43):24814–24829, 2019.

[603] Qiang Chang, Mohammad Ali Darabi, Yuqing Liu, Yunfan He, Wen Zhong, Kibret Mequanin, Bingyun Li, Feng Lu, and Malcolm M. Q. Xing. Hydrogels from natural egg white with extraordinary stretchability, direct-writing 3D printability and self-healing for fabrication of electronic sensors and actuators. *J. Mater. Chem. A*, 7(42):24626–24640, 2019.

[604] Kai Huang, Shaoming Dong, Jinshan Yang, Jingyi Yan, Yudong Xue, Xiao You, Jianbao Hu, Le Gao, Xiangyu Zhang, and Yusheng Ding. Three-dimensional printing of a tunable graphene-based elastomer for strain sensors with ultrahigh sensitivity. *Carbon*, 143:63–72, Mar 2019.

[605] Steve J. A. Majerus, Hao Chong, David Ariando, Connor Swingle, Joseph Potkay, Kath Bogie, and Christian A. Zorman. Vascular graft pressure-flow monitoring using 3D printed MWCNT-PDMS strain sensors. In *2018 40th annual international conference of the IEEE engineering in medicine and biology society (EMBC)*, pp. 2989–2992, Honolulu, HI, July 2018. IEEE.

[606] Qinghua Wu, Shibo Zou, Frédérick P. Gosselin, Daniel Therriault, and Marie-Claude Heuzey. 3D printing of a self-healing nanocomposite for stretchable sensors. *J. Mater. Chem. C*, 6(45):12180–12186, 2018.

[607] Ziya Wang, Xiao Guan, Huayi Huang, Haifei Wang, Waner Lin, and Zhengchun Peng. Full 3D printing of stretchable piezoresistive sensor with hierarchical porosity and multimodulus architecture. *Adv. Funct. Mater.*, 29(11):1807569, Mar 2019.

[608] Shuang-Zhuang Guo, Kaiyan Qiu, Fanben Meng, Sung Hyun Park, and Michael C. McAlpine. 3D printed stretchable tactile sensors. *Adv. Mater.*, 29(27):1701218, July 2017.

[609] Zhijie Zhu, Hyun Soo Park, and Michael C. McAlpine. 3D printed deformable sensors. *Sci. Adv.*, 6(25):eaba5575, June 2020.

[610] Kshama Parate, Sonal V. Rangnekar, Dapeng Jing, Deyny L. Mendivelso-Perez, Shaowei Ding, Ethan B. Secor, Emily A. Smith, Jesse M. Hostetter, Mark C. Hersam, and Jonathan C. Claussen. Aerosol-jet-printed graphene immunosensor for label-free cytokine monitoring in serum. *ACS Appl. Mater. Interfaces*, 12(7):8592–8603, Feb 2020.

[611] Robert Herbert, Saswat Mishra, Hyo-Ryoung Lim, Hyoungsuk Yoo, and Woon-Hong Yeo. Fully printed, wireless, stretchable implantable biosystem toward batteryless, real-time monitoring of cerebral aneurysm hemodynamics. *Adv. Sci.*, 6(18):1901034, Sept 2019.

[612] Robert Bogue. 3D printing: an emerging technology for sensor fabrication. *Sens. Rev.*, 36(4):333–338, Sept 2016.

[613] Xiangyang Jiao, Hui He, Wei Qian, Guanghua Li, Guangyao Shen, Xiao Li, Chong Ding, Douglas White, Stephen Scearce, Yaochao Yang, and David Pommerenke. Designing a 3-d printing-based channel emulator with printable electromagnetic materials. *IEEE Trans. Electromagn. Compat.*, 57(4):868–876, Aug 2015.

[614] A. Dorigato, V. Moretti, S. Dul, S. H. Unterberger, and A. Pegoretti. Electrically conductive nanocomposites for fused deposition modelling. *Synth. Met.*, 226:7–14, Apr 2017.

[615] H. H. Hamzah, O. Keattch, M. S. Yeoman, D. Covill, and B. A. Patel. Three-dimensional-printed electrochemical sensor for simultaneous dual monitoring of serotonin overflow and circular muscle contraction. *Anal. Chem.*, 91(18):12014–12020, Sept 2019.

[616] Vera Katic, Pãmyla L. dos Santos, Matheus F. dos Santos, Bruno M. Pires, Hugo C. Loureiro, Ana P. Lima, Júlia C. M. Queiroz, Richard Landers, Rodrigo A. A. Muñoz, and Juliano A. Bonacin. 3D printed graphene electrodes modified with Prussian blue: emerging electrochemical sensing platform for peroxide detection. *ACS Appl. Mater. Interfaces*, 11(38):35068–35078, Sept 2019.

[617] Muhammad Zafir Mohamad Nasir, Filip Novotný, Osamah Alduhaish, and Martin Pumera. 3D-printed electrodes for the detection of mycotoxins in food. *Electrochem. Commun.*, 115:106735, June 2020.

[618] Diego P. Rocha, André L. Squissato, Sarah M. da Silva, Eduardo M. Richter, and Rodrigo A. A. Munoz. Improved electrochemical detection of metals in biological samples using 3D-printed electrode: chemical/electrochemical treatment exposes carbon-black conductive sites. *Electrochim. Acta*, 335:135688, Mar 2020.

[619] Rafael M. Cardoso, Pablo R. L. Silva, Ana P. Lima, Diego P. Rocha, Thiago C. Oliveira, Thiago M. do Prado, Elson L. Fava, Orlando Fatibello-Filho, Eduardo M. Richter, and Rodrigo A. A. Muñoz. 3D-printed graphene/polylactic acid electrode for bioanalysis: biosensing of glucose and simultaneous determination of uric acid and nitrite in biological fluids. *Sens. Actuators B, Chem.*, 307:127621, Mar 2020.

[620] Gustavo Martins, Jeferson L. Gogola, Lucas H. Budni, Bruno C. Janegitz, Luiz H. Marcolino-Junior, and Márcio F. Bergamini. 3D-printed electrode as a new platform for electrochemical immunosensors for virus detection. *Anal. Chim. Acta*, 1147:30–37, Feb 2021.

[621] P. Salvo, R. Raedt, E. Carrette, D. Schaubroeck, J. Vanfleteren, and L. Cardon. A 3D printed dry electrode for ECG/EEG recording. *Sens. Actuators Phys.*, 174:96–102, Feb 2012.

[622] Josef F. Christ, Nahal Aliheidari, Amir Ameli, and Petra Pötschke. 3D printed highly elastic strain sensors of multiwalled carbon nanotube/thermoplastic polyurethane nanocomposites. *Mater. Des.*, 131:394–401, Oct 2017.

[623] Zhongming Li, Dong Feng, Bin Li, Delong Xie, and Yi Mei. FDM printed MXene/MnFe2O4/MWCNTs reinforced TPU composites with 3D Voronoi structure for sensor and electromagnetic shielding applications. *Compos. Sci. Technol.*, 231:109803, Jan 2023.

[624] Cheng-Kuan Su and Jo-Chin Chen. One-step three-dimensional printing of enzyme/substrate–incorporated devices for glucose testing. *Anal. Chim. Acta*, 1036:133–140, Dec 2018.

[625] Adaris M. López Marzo, Carmen C. Mayorga-Martinez, and Martin Pumera. 3D-printed graphene direct electron transfer enzyme biosensors. *Biosens. Bioelectron.*, 151:111980, Mar 2020.

[626] Ting Xiao, Cheng Qian, Ruixue Yin, Kemin Wang, Yang Gao, and Fuzhen Xuan. 3D printing of flexible strain sensor array based on uv-curable multiwalled carbon nanotube/elastomer composite. *Adv. Mater. Technol.*, 6(1):2000745, Jan 2021.

[627] Yang Yang, Xiangjia Li, Ming Chu, Haofan Sun, Jie Jin, Kunhao Yu, Qiming Wang, Qifa Zhou, and Yong Chen. Electrically assisted 3D printing of nacre-inspired structures with self-sensing capability. *Sci. Adv.*, 5(4):eaau9490, Apr 2019.

[628] Ryan L. Truby, Michael Wehner, Abigail K. Grosskopf, Daniel M. Vogt, Sebastien G. M. Uzel, Robert J. Wood, and Jennifer A. Lewis. Soft somatosensitive actuators via embedded 3D printing. *Adv. Mater.*, 30(15):1706383, Apr 2018.

[629] Morteza Amjadi and Metin Sitti. High-performance multiresponsive paper actuators. *ACS Nano*, 10(11):10202–10210, Nov 2016.

[630] Yoshio Ishiguro and Ivan Poupyrev. 3D printed interactive speakers. In *Proceedings of the SIGCHI conference on human factors in computing systems*, pp. 1733–1742, Toronto Ontario Canada, Apr 2014. ACM.

[631] Gilberto Siqueira, Dimitri Kokkinis, Rafael Libanori, Michael K. Hausmann, Amelia Sydney Gladman, Antonia Neels, Philippe Tingaut, Tanja Zimmermann, Jennifer A. Lewis, and André R. Studart. Cellulose nanocrystal inks for 3D printing of textured cellular architectures. *Adv. Funct. Mater.*, 27(12):1604619, Mar 2017.

[632] Dimitri Kokkinis, Manuel Schaffner, and André R. Studart. Multimaterial magnetically assisted 3D printing of composite materials. *Nat. Commun.*, 6(1):8643, Dec 2015.

[633] Eung Seok Park, Yenhao Chen, Tsu-Jae King Liu, and Vivek Subramanian. A new switching device for printed electronics: inkjet-printed microelectromechanical relay. *Nano Lett.*, 13(11):5355–5360, Nov 2013.

[634] Jeffrey Ian Lipton, Sarah Angle, and Hod Lipson. 3D printable wax-silicone actuators. In *25th International Solid Freeform Fabrication Symposium*, University of Texas at Austin, 2014. https://repositories.lib.utexas.edu/handle/2152/89228.

[635] Robert MacCurdy, Anthony McNicoll, and Hod Lipson. Bitblox: printable digital materials for electromechanical machines. *Int. J. Robot. Res.*, 33(10):1342–1360, Sept 2014.

[636] Writam Banerjee, Revannath Dnyandeo Nikam, and Hyunsang Hwang. Prospect and challenges of analog switching for neuromorphic hardware. *Appl. Phys. Lett.*, 120(6):060501, Feb 2022.

[637] Lu Lu, Ping Guo, and Yayue Pan. Magnetic-field-assisted projection stereolithography for three-dimensional printing of smart structures. *J. Manuf. Sci. Eng.*, 139(7):071008, July 2017.

[638] Matt Zarek, Michael Layani, Ido Cooperstein, Ela Sachyani, Daniel Cohn, and Shlomo Magdassi. 3D printing of shape memory polymers for flexible electronic devices. *Adv. Mater.*, 28(22):4449–4454, June 2016.

[639] Marcel Suter, Li Zhang, Erdem C. Siringil, Christian Peters, Tessa Luehmann, Olgac Ergeneman, Kathrin E. Peyer, Bradley J. Nelson, and Christofer Hierold. Superparamagnetic microrobots: fabrication by two-photon polymerization and biocompatibility. *Biomed. Microdevices*, 15(6):997–1003, Dec 2013.

[640] Petra S. Dittrich and Andreas Manz. Lab-on-a-chip: microfluidics in drug discovery. *Nat. Rev. Drug Discov.*, 5(3):210–218, Mar 2006.

[641] Johan U. Lind, Travis A. Busbee, Alexander D. Valentine, Francesco S. Pasqualini, Hongyan Yuan, Moran Yadid, Sung-Jin Park, Arda Kotikian, Alexander P. Nesmith, Patrick H. Campbell, Joost J. Vlassak, Jennifer A. Lewis, and Kevin K. Parker. Instrumented cardiac microphysiological devices via multimaterial three-dimensional printing. *Nat. Mater.*, 16(3):303–308, Mar 2017.

[642] Erik Trampe, Klaus Koren, Ashwini Rahul Akkineni, Christian Senwitz, Felix Krujatz, Anja Lode, Michael Gelinsky, and Michael Kühl. Functionalized bioink with optical sensor nanoparticles for O_2 imaging in 3D-bioprinted constructs. *Adv. Funct. Mater.*, 28(45):1804411, Nov 2018.

[643] Philip J. Kitson, Mali H. Rosnes, Victor Sans, Vincenza Dragone, and Leroy Cronin. Configurable 3D-printed millifluidic and microfluidic 'lab on a chip' reactionware devices. *Lab Chip*, 12(18):3267, 2012.

[644] Kyoung G. Lee, Kyun Joo Park, Seunghwan Seok, Sujeong Shin, Do Hyun Kim, Jung Youn Park, Yun Seok Heo, Seok Jae Lee, and Tae Jae Lee. 3D printed modules for integrated microfluidic devices. *RSC Adv.*, 4(62):32876–32880, 2014.

[645] Wonjae Lee, Donghoon Kwon, Woong Choi, Gyoo Yeol Jung, Anthony K. Au, Albert Folch, and Sangmin Jeon. 3D-printed microfluidic device for the detection of pathogenic bacteria using size-based separation in helical channel with trapezoid cross-section. *Sci. Rep.*, 5(1):7717, July 2015.

[646] Bing Li, Haijie Tan, Salzitsa Anastasova, Maura Power, Florent Seichepine, and Guang-Zhong Yang. A bio-inspired 3D micro-structure for graphene-based bacteria sensing. *Biosens. Bioelectron.*, 123:77–84, Jan 2019.

[647] Yasumitsu Miyata, Kazunari Shiozawa, Yuki Asada, Yutaka Ohno, Ryo Kitaura, Takashi Mizutani, and Hisanori Shinohara. Length-sorted semiconducting carbon nanotubes for high-mobility thin film transistors. *Nano Res.*, 4(10):963–970, Oct 2011.

[648] Dong-ming Sun, Marina Y. Timmermans, Ying Tian, Albert G. Nasibulin, Esko I. Kauppinen, Shigeru Kishimoto, Takashi Mizutani, and Yutaka Ohno. Flexible high-performance carbon nanotube integrated circuits. *Nat. Nanotechnol.*, 6(3):156–161, Mar 2011.

[649] Dheeraj Jain, Nima Rouhi, Christopher Rutherglen, Crystal G. Densmore, Stephen K. Doorn, and Peter J. Burke. Effect of source, surfactant, and deposition process on electronic properties of nanotube arrays. *J. Nanomater.*, 2011:1–7, 2011.

[650] Wei Gao, Yunbo Zhang, Devarajan Ramanujan, Karthik Ramani, Yong Chen, Christopher B. Williams, Charlie C. L. Wang, Yung C. Shin, Song Zhang, and Pablo D. Zavattieri. The status, challenges, and future of additive manufacturing in engineering. *Comput. Aided Des.*, 69:65–89, Dec 2015.

[651] Yong-Young Noh, Ni Zhao, Mario Caironi, and Henning Sirringhaus. Downscaling of self-aligned, all-printed polymer thin-film transistors. *Nat. Nanotechnol.*, 2(12):784–789, Dec 2007.

[652] Masayuki Abe. Development of submicron resolution R2R printing process for PE sensors. Orlando, FL, Dec 2015.

[653] Yang-Seok Park, Junyoung Kim, Jung Min Oh, Seungyoung Park, Seungse Cho, Hyunhyub Ko, and Yoon-Kyoung Cho. Near-field electrospinning for three-dimensional stacked nanoarchitectures with high aspect ratios. *Nano Lett.*, 20(1):441–448, Jan 2020.

[654] Julian Schneider, Patrik Rohner, Deepankur Thureja, Martin Schmid, Patrick Galliker, and Dimos Poulikakos. Electrohydrodynamic nanodrip printing of high aspect ratio metal grid transparent electrodes. *Adv. Funct. Mater.*, 26(6):833–840, Feb 2016.

[655] Enrico Sowade, Maxim Polomoshnov, Andreas Willert, and Reinhard R. Baumann. Toward 3D-printed electronics: inkjet-printed vertical metal wire interconnects and screen-printed batteries. *Adv. Eng. Mater.*, 21(10):1900568, Oct 2019.

[656] Mike Judd and Keith Brindley. *Soldering in electronics assembly*. Elsevier, Mar 1999.

[657] Konrad Kiełbasiński, Jerzy Szałapak, Małgorzata Jakubowska, Anna Młoźniak, Elżbieta Zwierkowska, Jakub Krzemiński, and Marian Teodorczyk. Influence of nanoparticles content in silver paste on mechanical and electrical properties of LTJT joints. *Adv. Powder Technol.*, 26(3):907–913, May 2015.

[658] Kim S. Siow. Mechanical properties of nano-silver joints as die attach materials. *J. Alloys Compd.*, 514:6–19, Feb 2012.

[659] R. Blanco and V. Mansingh. Thermal characteristics of buried resistors. In *Fifth annual IEEE semiconductor thermal and temperature measurement symposium*, pp. 25–29, San Diego, CA, USA, 1989. IEEE.

[660] C. L. Chen, C. K. Chen, J. A. Burns, D.-R. Yost, K. Warner, J. M. Knecht, P. W. Wyatt, D. A. Shibles, and C. L. Keast. Thermal effects of three dimensional integrated circuit stacks. In *2007 IEEE international SOI conference*, pp. 91–92, Indian Wells, CA, USA, Oct 2007. IEEE.

[661] Eveliina Juntunen, Vishal Gandhi, Satu Ylimaula, and Arttu Huttunen. Improving the performance of advertising led displays by in-mould integration. In *2018 IMAPS nordic conference on microelectronics packaging (NordPac)*, pp. 45–49, Oulu, June 2018. IEEE.

[662] Hessel H. H. Maalderink, Fabien B. J. Bruning, Mathijs M. R. De Schipper, John J. J. Van Der Werff, Wijnand W. C. Germs, Joris J. C. Remmers, and Erwin R. Meinders. 3D printed structural electronics: embedding and connecting electronic components into freeform electronic devices. *Plast. Rubber Compos.*, 47(1):35–41, Jan 2018.

[663] Shaoyong Yang, Dawei Xiang, Angus Bryant, Philip Mawby, Li Ran, and Peter Tavner. Condition monitoring for device reliability in power electronic converters: a review. *IEEE Trans. Power Electron.*, 25(11):2734–2752, Nov 2010.

[664] S. Suresh. *Fatigue of materials*. Cambridge University Press, Cambridge; New York, 2nd edition, 1998.

[665] Bartłomiej Podsiadły, Andrzej Skalski, and Marcin Słoma. Soldering of electronics components on 3D-printed conductive substrates. *Materials*, 14(14):3850, July 2021.

[666] Jie Hu and Min-Feng Yu. Meniscus-confined three-dimensional electrodeposition for direct writing of wire bonds. *Science*, 329(5989):313–316, July 2010.

[667] Bok Y. Ahn, Eric B. Duoss, Michael J. Motala, Xiaoying Guo, Sang-Il Park, Yujie Xiong, Jongseung Yoon, Ralph G. Nuzzo, John A. Rogers, and Jennifer A. Lewis. Omnidirectional printing of flexible, stretchable, and spanning silver microelectrodes. *Science*, 323(5921):1590–1593, Mar 2009.

[668] Mark A. Skylar-Scott, Jochen Mueller, Claas W. Visser, and Jennifer A. Lewis. Voxelated soft matter via multimaterial multinozzle 3D printing. *Nature*, 575(7782):330–335, Nov 2019.

[669] Eshwar Cholleti, Jonathan Stringer, Mahtab Assadian, Virginie Battmann, Chris Bowen, and Kean Aw. Highly stretchable capacitive sensor with printed carbon black electrodes on barium titanate elastomer composite. *Sensors*, 19(1):42, Dec 2018.

[670] Zhouyue Lei, Quankang Wang, and Peiyi Wu. A multifunctional skin-like sensor based on a 3D printed thermo-responsive hydrogel. *Mater. Horiz.*, 4(4):694–700, 2017.

[671] Ali Gökhan Demir and Barbara Previtali. Multi-material selective laser melting of Fe/Al-12Si components. *Manuf. Lett.*, 11:8–11, Jan 2017.

[672] Qi Ge, Amir Hosein Sakhaei, Howon Lee, Conner K. Dunn, Nicholas X. Fang, and Martin L. Dunn. Multimaterial 4D printing with tailorable shape memory polymers. *Sci. Rep.*, 6(1):31110, Nov 2016.

[673] Z. H. Liu, D. Q. Zhang, S. L. Sing, C. K. Chua, and L. E. Loh. Interfacial characterization of SLM parts in multi-material processing: metallurgical diffusion between 316L stainless steel and C18400 copper alloy. *Mater. Charact.*, 94:116–125, Aug 2014.

[674] Xiang-Yu Yin, Yue Zhang, Junfeng Xiao, Carolyn Moorlag, and Jun Yang. Monolithic dual-material 3D printing of ionic skins with long-term performance stability. *Adv. Funct. Mater.*, 29(39):1904716, Sept 2019.

[675] Chi Zhou, Yong Chen, Zhigang Yang, and Behrokh Khoshnevis. Development of a multi-material mask-image-projection-based stereolithography for the fabrication of digital materials. In *Solid freeform fabrication symposium*, 2011.

[676] Daehoon Han, Chen Yang, Nicholas X. Fang, and Howon Lee. Rapid multi-material 3D printing with projection micro-stereolithography using dynamic fluidic control. *Addit. Manuf.*, 27:606–615, May 2019.

[677] Mehrshad Mehrpouya, Daniel Tuma, Tom Vaneker, Mohamadreza Afrasiabi, Markus Bambach, and Ian Gibson. Multimaterial powder bed fusion techniques. *RPJ*, 28(11):1–19, Dec 2022.

[678] Zixiang Weng, Yu Zhou, Wenxiong Lin, T. Senthil, and Lixin Wu. Structure-property relationship of nano enhanced stereolithography resin for desktop SLA 3D printer. *Composites, Part A, Appl. Sci. Manuf.*, 88:234–242, Sept 2016.

[679] Michael Layani, Ido Cooperstein, and Shlomo Magdassi. UV crosslinkable emulsions with silver nanoparticles for inkjet printing of conductive 3D structures. *J. Mater. Chem. C*, 1(19):3244, 2013.

[680] Ehab Saleh, Fan Zhang, Yinfeng He, Jayasheelan Vaithilingam, Javier Ledesma Fernandez, Ricky Wildman, Ian Ashcroft, Richard Hague, Phill Dickens, and Christopher Tuck. 3D inkjet printing of electronics using uv conversion. *Adv. Mater. Technol.*, 2(10):1700134, Oct 2017.

[681] Bartlomiej Walpuski and Marcin Słoma. Manufacturing methods and challenges in structural electronics. In Ryszard S. Romaniuk and Maciej Linczuk, editors, *Photonics applications in astronomy, communications, industry, and high-energy physics experiments 2019*, p. 177, Wilga, Poland, Nov 2019. SPIE.

[682] Jeevan Persad and Sean Rocke. Multi-material 3D printed electronic assemblies: a review. *Results Eng.*, 16:100730, Dec 2022.

[683] Florens Wasserfall. Topology-aware routing of electric wires in FDM-printed objects. In *Solid freeform fabrication symposium*, 2018.

[684] Faez Alkadi, Kyung-Chang Lee, Abdullateef H. Bashiri, and Jae-Won Choi. Conformal additive manufacturing using a direct-print process. *Addit. Manuf.*, 32:100975, Mar 2020.

[685] Jorge G. Cham, Beth L. Pruitt, Mark R. Cutkosky, Mike Binnard, Lee E. Weiss, and Gennady Neplotnik. Layered manufacturing with embedded components: process planning considerations. In *Volume 4: 4th design for manufacturing conference*, pp. 93–101, Las Vegas, Nevada, USA, Sep 1999. American Society of Mechanical Engineers.

[686] Zhijie Zhu, Shuang-Zhuang Guo, Tessa Hirdler, Cindy Eide, Xiaoxiao Fan, Jakub Tolar, and Michael C. McAlpine. 3D printed functional and biological materials on moving freeform surfaces. *Adv. Mater.*, 30(23):1707495, June 2018.

[687] nScrypt. nScrypt 3Dn series. www.nscrypt.com.

[688] Devin J. Roach, Craig M. Hamel, Conner K. Dunn, Marshall V. Johnson, Xiao Kuang, and H. Jerry Qi. The m4 3D printer: a multi-material multi-method additive manufacturing platform for future 3D printed structures. *Addit. Manuf.*, 29:100819, Oct 2019.

[689] Neotech AMT. Neotech AMT 15x BT. www.neotech-amt.com.

[690] Petri Kanninen, Christoffer Johans, Juha Merta, and Kyösti Kontturi. Influence of ligand structure on the stability and oxidation of copper nanoparticles. *J. Colloid Interface Sci.*, 318(1):88–95, Feb 2008.

[691] Shlomo Magdassi, Michael Grouchko, and Alexander Kamyshny. Copper nanoparticles for printed electronics: routes towards achieving oxidation stability. *Materials*, 3(9):4626–4638, Sept 2010.

[692] Xiaofeng Dai, Wen Xu, Teng Zhang, Hongbin Shi, and Tao Wang. Room temperature sintering of Cu-Ag core-shell nanoparticles conductive inks for printed electronics. *Chem. Eng. J.*, 364:310–319, May 2019.

[693] Anna Pajor-Świerzy, Krzysztof Szczepanowicz, Alexander Kamyshny, and Shlomo Magdassi. Metallic core-shell nanoparticles for conductive coatings and printing. *Adv. Colloid Interface Sci.*, 299:102578, Jan 2022.

[694] Changsoo Lee, Na Rae Kim, Jahyun Koo, Yung Jong Lee, and Hyuck Mo Lee. Cu-Ag core–shell nanoparticles with enhanced oxidation stability for printed electronics. *Nanotechnology*, 26(45):455601, Nov 2015.

[695] Chang Kyu Kim, Gyoung-Ja Lee, Min Ku Lee, and Chang Kyu Rhee. A novel method to prepare Cu@Ag core–shell nanoparticles for printed flexible electronics. *Powder Technol.*, 263:1–6, Sept 2014.

[696] Tae Gon Kim, Hye Jin Park, Kyoohee Woo, Sunho Jeong, Youngmin Choi, and Su Yeon Lee. Enhanced oxidation-resistant Cu@Ni core–shell nanoparticles for printed flexible electrodes. *ACS Appl. Mater. Interfaces*, 10(1):1059–1066, Jan 2018.

[697] Yun Hwan Jo, Inyu Jung, Na Rae Kim, and Hyuck Mo Lee. Synthesis and characterization of highly conductive Sn–Ag bimetallic nanoparticles for printed electronics. *J. Nanopart. Res.*, 14(4):782, Mar 2012.

[698] T. D. Papathanasiou and Andre Benard. *Flow-induced alignment in composite materials*. Woodhead publishing series in composites science and engineering. Woodhead Publishing, Duxford Cambridge, MA Kidlington, 2nd edition, 2022.

[699] Zuojia Wang, Feng Cheng, Thomas Winsor, and Yongmin Liu. Optical chiral metamaterials: a review of the fundamentals, fabrication methods and applications. *Nanotechnology*, 27(41):412001, Oct 2016.

[700] Benny Bar-On, Xiaomeng Sui, Konstantin Livanov, Ben Achrai, Estelle Kalfon-Cohen, Erica Wiesel, and H. Daniel Wagner. Structural origins of morphing in plant tissues. *Appl. Phys. Lett.*, 105(3):033703, July 2014.

[701] Clara Fuciños, Pablo Fuciños, Martín Míguez, Issa Katime, Lorenzo M. Pastrana, and María L. Rúa. Temperature- and pH-sensitive nanohydrogels of poly(n-isopropylacrylamide) for food packaging applications: modelling the swelling-collapse behaviour. *PLoS ONE*, 9(2):e87190, Feb 2014.

[702] A. Sydney Gladman, Elisabetta A. Matsumoto, Ralph G. Nuzzo, L. Mahadevan, and Jennifer A. Lewis. Biomimetic 4D printing. *Nat. Mater.*, 15(4):413–418, Apr 2016.

[703] Yifei Jin, Chengcheng Liu, Wenxuan Chai, Ashley Compaan, and Yong Huang. Self-supporting nanoclay as internal scaffold material for direct printing of soft hydrogel composite structures in air. *ACS Appl. Mater. Interfaces*, 9(20):17456–17465, May 2017.

[704] Yanyang Han, Chee Chuan J. Yeo, Dairong Chen, Fei Wang, Yiting Chong, Xu Li, Xiuling Jiao, and FuKe Wang. Nanowire enhanced dimensional accuracy in acrylate resin-based 3D printing. *New J. Chem.*, 41(16):8407–8412, 2017.

[705] Alireza Nafari and Henry A Sodano. Electromechanical modeling and experimental verification of a direct write nanocomposite. *Smart Mater. Struct.*, 28(4):045014, Apr 2019.

[706] J. William Boley, Kundan Chaudhary, Thomas J. Ober, Mohammadreza Khorasaninejad, Wei Ting Chen, Erik Hanson, Ashish Kulkarni, Jaewon Oh, Jinwoo Kim, Larry K. Aagesen, Alexander Y. Zhu, Federico Capasso, Katsuyo Thornton, Paul V. Braun, and Jennifer A. Lewis. High-operating-temperature direct ink writing of mesoscale eutectic architectures. *Adv. Mater.*, 29(7):1604778, Feb 2017.

[707] Chao Wang, Xusheng Wang, Yanjie Wang, Jitao Chen, Henghui Zhou, and Yunhui Huang. Macroporous free-standing nano-sulfur/reduced graphene oxide paper as stable cathode for lithium-sulfur battery. *Nano Energy*, 11:678–686, Jan 2015.

[708] Minxuan Kuang, Jingxia Wang, Bin Bao, Fengyu Li, Libin Wang, Lei Jiang, and Yanlin Song. Inkjet printing patterned photonic crystal domes for wide viewing-angle displays by controlling the sliding three phase contact line. *Adv. Opt. Mater.*, 2(1):34–38, Jan 2014.

[709] Meng Su, Zhandong Huang, Yifan Li, Xin Qian, Zheng Li, Xiaotian Hu, Qi Pan, Fengyu Li, Lihong Li, and Yanlin Song. A 3D self-shaping strategy for nanoresolution multicomponent architectures. *Adv. Mater.*, 30(3):1703963, Jan 2018.

[710] C. R. P. Courtney, C.-K. Ong, B. W. Drinkwater, A. L. Bernassau, P. D. Wilcox, and D. R. S. Cumming. Manipulation of particles in two dimensions using phase controllable ultrasonic standing waves. *Proc. R. Soc. A, Math. Phys. Eng. Sci.*, 468(2138):337–360, Feb 2012.

[711] J. Greenhall, F. Guevara Vasquez, and B. Raeymaekers. Ultrasound directed self-assembly of user-specified patterns of nanoparticles dispersed in a fluid medium. *Appl. Phys. Lett.*, 108(10):103103, Mar 2016.

[712] Lu Lu, Xiaohui Tang, Shan Hu, and Yayue Pan. Acoustic field-assisted particle patterning for smart polymer composite fabrication in stereolithography. *3D Print. Addit. Manuf.*, 5(2):151–159, June 2018.

[713] Soheila Shabaniverki, Sarah Thorud, and Jaime J. Juárez. Vibrationally directed assembly of micro- and nanoparticle-polymer composites. *Chem. Eng. Sci.*, 192:1209–1217, Dec 2018.

[714] Doruk Erdem Yunus, Salman Sohrabi, Ran He, Wentao Shi, and Yaling Liu. Acoustic patterning for 3D embedded electrically conductive wire in stereolithography. *J. Micromech. Microeng.*, 27(4):045016, Apr 2017.

[715] Luis A. Chavez, Jaime E. Regis, Luis C. Delfin, Carlos A. Garcia Rosales, Hoejin Kim, Norman Love, Yingtao Liu, and Yirong Lin. Electrical and mechanical tuning of 3D printed photopolymer–MWCNT nanocomposites through in situ dispersion. *J. Appl. Polym. Sci.*, 136(22):47600, June 2019.

[716] Horacio D. Espinosa, Allison L. Juster, Felix J. Latourte, Owen Y. Loh, David Gregoire, and Pablo D. Zavattieri. Tablet-level origin of toughening in abalone shells and translation to synthetic composite materials. *Nat. Commun.*, 2(1):173, Feb 2011.

[717] Yang Yang, Zeyu Chen, Xuan Song, Zhuofeng Zhang, Jun Zhang, K. Kirk Shung, Qifa Zhou, and Yong Chen. Biomimetic anisotropic reinforcement architectures by electrically assisted nanocomposite 3D printing. *Adv. Mater.*, 29(11):1605750, Mar 2017.

[718] Takeshi Nakamoto and Sho Marukado. Properties of photopolymer part with aligned short ferromagnetic fibers. *Int. J. Autom. Technol.*, 10(6):916–922, Nov 2016.

[719] Bethany C. Gross, Jayda L. Erkal, Sarah Y. Lockwood, Chengpeng Chen, and Dana M. Spence. Evaluation of 3D printing and its potential impact on biotechnology and the chemical sciences. *Anal. Chem.*, 86(7):3240–3253, Apr 2014.

[720] Samuel Hales, Eric Tokita, Rajan Neupane, Udayan Ghosh, Brian Elder, Douglas Wirthlin, and Yong Lin Kong. 3D printed nanomaterial-based electronic, biomedical, and bioelectronic devices. *Nanotechnology*, 31(17):172001, Apr 2020.

[721] Yong Lin Kong, Maneesh K. Gupta, Blake N. Johnson, and Michael C. McAlpine. 3D printed bionic nanodevices. *Nano Today*, 11(3):330–350, June 2016.

[722] David Chimene, Kimberly K. Lennox, Roland R. Kaunas, and Akhilesh K. Gaharwar. Advanced bioinks for 3D printing: a materials science perspective. *Ann. Biomed. Eng.*, 44(6):2090–2102, June 2016.

[723] Se-Jun Lee, Timothy Esworthy, Seth Stake, Shida Miao, Yi Y. Zuo, Brent T. Harris, and Lijie Grace Zhang. Advances in 3D bioprinting for neural tissue engineering. *Adv. Biosyst.*, 2(4):1700213, Apr 2018.

[724] Wen Jiang, Hu Li, Zhuo Liu, Zhe Li, Jingjing Tian, Bojing Shi, Yang Zou, Han Ouyang, Chaochao Zhao, Luming Zhao, Rong Sun, Hairong Zheng, Yubo Fan, Zhong Lin Wang, and Zhou Li. Fully bioabsorbable natural-materials-based triboelectric nanogenerators. *Adv. Mater.*, 30(32):1801895, Aug 2018.

[725] Hyunwoo Yuk, Claudia E. Varela, Christoph S. Nabzdyk, Xinyu Mao, Robert F. Padera, Ellen T. Roche, and Xuanhe Zhao. Dry double-sided tape for adhesion of wet tissues and devices. *Nature*, 575(7781):169–174, Nov 2019.

[726] Youngsik Lee, Jaemin Kim, Ja Hoon Koo, Tae-Ho Kim, and Dae-Hyeong Kim. Nanomaterials for bioelectronics and integrated medical systems. *Korean J. Chem. Eng.*, 35(1):1–11, Jan 2018.

[727] Yuxin Liu, Jia Liu, Shucheng Chen, Ting Lei, Yeongin Kim, Simiao Niu, Huiliang Wang, Xiao Wang, Amir M. Foudeh, Jeffrey B.-H. Tok, and Zhenan Bao. Soft and elastic hydrogel-based microelectronics for localized low-voltage neuromodulation. *Nat. Biomed. Eng.*, 3(1):58–68, Jan 2019.

[728] Zhifeng Shi, Faming Zheng, Zhitao Zhou, Meng Li, Zhen Fan, Huanpeng Ye, Shan Zhang, Ting Xiao, Liang Chen, Tiger H. Tao, Yun-Lu Sun, and Ying Mao. Silk-enabled conformal multifunctional bioelectronics for investigation of spatiotemporal epileptiform activities and multimodal neural encoding/decoding. *Adv. Sci.*, 6(9):1801617, May 2019.

[729] Eli J. Curry, Thinh T. Le, Ritopa Das, Kai Ke, Elise M. Santorella, Debayon Paul, Meysam T. Chorsi, Khanh T. M. Tran, Jeffrey Baroody, Emily R. Borges, Brian Ko, Asiyeh Golabchi, Xiaonan Xin, David Rowe, Lixia Yue, Jianlin Feng, M. Daniela Morales-Acosta, Qian Wu, I-Ping Chen, X. Tracy Cui, Joel Pachter, and Thanh D. Nguyen. Biodegradable nanofiber-based piezoelectric transducer. *Proc. Natl. Acad. Sci.*, 117(1):214–220, Jan 2020.

[730] Clementine M. Boutry, Levent Beker, Yukitoshi Kaizawa, Christopher Vassos, Helen Tran, Allison C. Hinckley, Raphael Pfattner, Simiao Niu, Junheng Li, Jean Claverie, Zhen Wang, James Chang, Paige M. Fox, and Zhenan Bao. Biodegradable and flexible arterial-pulse sensor for the wireless monitoring of blood flow. *Nat. Biomed. Eng.*, 3(1):47–57, Jan 2019.

[731] Manu S. Mannoor, Ziwen Jiang, Teena James, Yong Lin Kong, Karen A. Malatesta, Winston O. Soboyejo, Naveen Verma, David H. Gracias, and Michael C. McAlpine. 3D printed bionic ears. *Nano Lett.*, 13(6):2634–2639, June 2013.

[732] Caokun Yang, Yong Xiang, Bin Liao, and Xiaoran Hu. 3D-printed bionic ear for sound identification and localization based on in situ polling of PVDF-TrFE film. *Macromol. Biosci.*, 23(2):2200374, Feb 2023.

[733] Moonchul Park, Sayemul Islam, Hye-Eun Kim, Jonathan Korostoff, Markus B. Blatz, Geelsu Hwang, and Albert Kim. Human oral motion-powered smart dental implant (SDI) for in situ ambulatory photo-biomodulation therapy. *Adv. Healthc. Mater.*, 9(16):2000658, Aug 2020.

[734] Mein Jin Tan, Cally Owh, Pei Lin Chee, Aung Ko Ko Kyaw, Dan Kai, and Xian Jun Loh. Biodegradable electronics: cornerstone for sustainable electronics and transient applications. *J. Mater. Chem. C*, 4(24):5531–5558, 2016.

[735] Huanyu Cheng and Vikas Vepachedu. Recent development of transient electronics. *Theor. Appl. Mech. Lett.*, 6(1):21–31, Jan 2016.

[736] Kun Kelvin Fu, Zhengyang Wang, Jiaqi Dai, Marcus Carter, and Liangbing Hu. Transient electronics: materials and devices. *Chem. Mater.*, 28(11):3527–3539, June 2016.

[737] Mihai Irimia-Vladu, Eric D. Głowacki, Gundula Voss, Siegfried Bauer, and Niyazi Serdar Sariciftci. Green and biodegradable electronics. *Mater. Today*, 15(7–8):340–346, July 2012.

[738] R. Anandkumar and S. Ramesh Babu. FDM filaments with unique segmentation since evolution: a critical review. *Prog. Addit. Manuf.*, 4(2):185–193, June 2019.

[739] R. Prabhu and A. Devaraju. Recent review of tribology, rheology of biodegradable and FDM compatible polymers. *Mater. Today Proc.*, 39:781–788, 2021.

[740] S. J. Eichhorn, A. Etale, J. Wang, L. A. Berglund, Y. Li, Y. Cai, C. Chen, E. D. Cranston, M. A. Johns, Z. Fang, G. Li, L. Hu, M. Khandelwal, K.-Y. Lee, K. Oksman, S. Pinitsoontorn, F. Quero, A. Sebastian, M. M. Titirici, Z. Xu, S. Vignolini, and B. Frka-Petesic. Current international research into cellulose as a functional nanomaterial for advanced applications. *J. Mater. Sci.*, 57(10):5697–5767, Mar 2022.

[741] Xiaohong Lan, Wenjian Li, Chongnan Ye, Laura Boetje, Théophile Pelras, Fitrilia Silvianti, Qi Chen, Yutao Pei, and Katja Loos. Scalable and degradable dextrin-based elastomers for wearable touch sensing. *ACS Appl. Mater. Interfaces*, acsami.2c15634, Dec 2022.

[742] Sherin Peter, Nathalie Lyczko, Deepu Gopakumar, Hanna J. Maria, Ange Nzihou, and Sabu Thomas. Nanocellulose and its derivative materials for energy and environmental applications. *J. Mater. Sci.*, 57(13):6835–6880, Apr 2022.

[743] Farsa Ram, Prashant Yadav, and Kadhiravan Shanmuganathan. Nanocellulose/melanin-based composites for energy, environment, and biological applications. *J. Mater. Sci.*, 57(30):14188–14216, Aug 2022.

[744] Meiling Wu, Ye Zhang, Lin Xu, Chunpeng Yang, Min Hong, Mingjin Cui, Bryson C. Clifford, Shuaiming He, Shuangshuang Jing, Yan Yao, and Liangbing Hu. A sustainable chitosan-zinc electrolyte for high-rate zinc-metal batteries. *Matter*, 5(10):3402–3416, Oct 2022.

[745] Jialin Bi, Zhangyin Yan, Lei Hao, Ashraf Y. Elnaggar, Salah M. El-Bahy, Fuhao Zhang, Islam H. El Azab, Qian Shao, Gaber A. M. Mersal, Junxiang Wang, Mina Huang, and Zhanhu Guo. Improving water resistance and mechanical properties of waterborne acrylic resin modified by octafluoropentyl methacrylate. *J. Mater. Sci.*, 58(3):1452–1464, Jan 2023.

[746] Christian Capello, Ulrich Fischer, and Konrad Hungerbühler. What is a green solvent? A comprehensive framework for the environmental assessment of solvents. *Green Chem.*, 9(9):927, 2007.

[747] Valerian E. Kagan, Nagarjun V. Konduru, Weihong Feng, Brett L. Allen, Jennifer Conroy, Yuri Volkov, Irina I. Vlasova, Natalia A. Belikova, Naveena Yanamala, Alexander Kapralov, Yulia Y. Tyurina, Jingwen Shi, Elena R. Kisin, Ashley R. Murray, Jonathan Franks, Donna Stolz, Pingping Gou, Judith Klein-Seetharaman, Bengt Fadeel, Alexander Star, and Anna A. Shvedova. Carbon nanotubes degraded by neutrophil myeloperoxidase induce less pulmonary inflammation. *Nat. Nanotechnol.*, 5(5):354–359, May 2010.

[748] Hector Villagarcia, Enkeleda Dervishi, Kanishka de Silva, Alexandru S. Biris, and Mariya V. Khodakovskaya. Surface chemistry of carbon nanotubes impacts the growth and expression of water channel protein in tomato plants. *Small*, 8(15):2328–2334, Aug 2012.

[749] M. Hilder, B. Winther-Jensen, and N. B. Clark. Paper-based, printed zinc–air battery. *J. Power Sources*, 194(2):1135–1141, Dec 2009.

[750] Amin Zareei, Vidhya Selvamani, Sarath Gopalakrishnan, Sachin Kadian, Murali Kannan Maruthamuthu, Zihao He, Juliane Nguyen, Haiyan Wang, and Rahim Rahimi. A biodegradable hybrid micro/nano conductive zinc paste for paper-based flexible bioelectronics. *Adv. Mater. Technol.*, 7(10):2101722, Oct 2022.

[751] Suk-Won Hwang, Xian Huang, Jung-Hun Seo, Jun-Kyul Song, Stanley Kim, Sami Hage-Ali, Hyun-Joong Chung, Hu Tao, Fiorenzo G. Omenetto, Zhenqiang Ma, and John A. Rogers. Materials for bioresorbable radio frequency electronics. *Adv. Mater.*, 25(26):3526–3531, July 2013.

[752] Lan Yin, Xian Huang, Hangxun Xu, Yanfeng Zhang, Jasper Lam, Jianjun Cheng, and John A. Rogers. Materials, designs, and operational characteristics for fully biodegradable primary batteries. *Adv. Mater.*, 26(23):3879–3884, June 2014.

[753] Seung-Kyun Kang, Jahyun Koo, Yoon Kyeung Lee, and John A. Rogers. Advanced materials and devices for bioresorbable electronics. *Acc. Chem. Res.*, 51(5):988–998, May 2018.

[754] Ivan K. Ilic, Leonardo Lamanna, Daniele Cortecchia, Pietro Cataldi, Alessandro Luzio, and Mario Caironi. Self-powered edible defrosting sensor. *ACS Sens.*, 7(10):2995–3005, Oct 2022.

[755] Mihai Irimia-Vladu, Pavel A. Troshin, Melanie Reisinger, Lyuba Shmygleva, Yasin Kanbur, Günther Schwabegger, Marius Bodea, Reinhard Schwödiauer, Alexander Mumyatov, Jeffrey W. Fergus, Vladimir F. Razumov, Helmut Sitter, Niyazi Serdar Sariciftci, and Siegfried Bauer. Biocompatible and biodegradable materials for organic field-effect transistors. *Adv. Funct. Mater.*, 20(23):4069–4076, Dec 2010.

[756] Christophe Collet, Alankar A. Vaidya, Marc Gaugler, Mark West, and Gareth Lloyd-Jones. Extrusion of PHA-containing bacterial biomass and the fate of endotoxins: a cost-reducing platform for applications in molding, coating and 3D printing. *Mater. Today Commun.*, 33:104162, Dec 2022.

[757] S. M. Lebedev, O. S. Gefle, E. T. Amitov, D. V. Zhuravlev, D. Y. Berchuk, and E. A. Mikutskiy. Mechanical properties of PLA-based composites for fused deposition modeling technology. *Int. J. Adv. Manuf. Technol.*, 97(1–4):511–518, July 2018.

[758] Razieh Hashemi Sanatgar, Aurélie Cayla, Christine Campagne, and Vincent Nierstrasz. Morphological and electrical characterization of conductive polylactic acid based nanocomposite before and after FDM 3D printing. *J. Appl. Polym. Sci.*, 136(6):47040, Feb 2019.

[759] Kimball Andersen, Yue Dong, and Woo Soo Kim. Highly conductive three-dimensional printing with low-melting metal alloy filament. *Adv. Eng. Mater.*, 19(11):1700301, Nov 2017.

[760] Jorge Mireles, David Espalin, David Roberson, Bob Zinniel, Francisco Medina, and Ryan Wicker. Fused deposition modeling of metals. In *Proc. solid free. fabr. symp. Austin TX USA*, pp. 6–8, 2012.

[761] Nirupama Warrier and Kunal H. Kate. Fused filament fabrication 3D printing with low-melt alloys. *Prog. Addit. Manuf.*, 3(1–2):51–63, June 2018.

[762] Yongze Yu, Fujun Liu, and Jing Liu. Direct 3D printing of low melting point alloy via adhesion mechanism. *Rapid Prototyping J.*, 23(3):642–650, Apr 2017.

[763] Adrian Bowyer. First reprapped circuit. *RepRap Blog*, Apr 2009. http://blog.reprap.org/search/label/reprapped%20electronics.

[764] Jorge Mireles, Ho-Chan Kim, In Hwan Lee, David Espalin, Francisco Medina, Eric MacDonald, and Ryan Wicker. Development of a fused deposition modeling system for low melting temperature metal alloys. *J. Electron. Packag.*, 135(1):011008, Feb 2013.

[765] Pengju Zhang, Yang Yu, Bowei Chen, Wei Wang, Sijian Wei, Wei Rao, and Qian Wang. Fast fabrication of double-layer printed circuits using bismuth-based low-melting alloy beads. *J. Mater. Chem. C*, 8(24):8028–8035, 2020.

[766] Ping Ren and Jingyan Dong. Direct fabrication of via interconnects by electrohydrodynamic printing for multi-layer 3D flexible and stretchable electronics. *Adv. Mater. Technol.*, 6(9):2100280, Sept 2021.

[767] Björn Gerdes. *StarJet printheads for printing molten solder jets at 320 °C and molten aluminum alloy droplets at 950 °C*. PhD thesis, Albert-Ludwigs-Universität Freiburg, 2019. https://freidok.uni-freiburg.de/data/150324, DOI: https://doi.org/10.6094/UNIFR/150324.

[768] Uranbileg Daalkhaijav, Osman Dogan Yirmibesoglu, Stephanie Walker, and Yigit Mengüç. Rheological modification of liquid metal for additive manufacturing of stretchable electronics. *Adv. Mater. Technol.*, 3(4):1700351, Apr 2018.

[769] Michael D. Dickey. Stretchable and soft electronics using liquid metals. *Adv. Mater.*, 29(27):1606425, July 2017.

[770] Young-Geun Park, Ga-Yeon Lee, Jiuk Jang, Su Min Yun, Enji Kim, and Jang-Ung Park. Liquid metal-based soft electronics for wearable healthcare. *Adv. Healthc. Mater.*, 10(17):2002280, Sept 2021.

[771] Qian Wang, Yang Yu, and Jing Liu. Preparations, characteristics and applications of the functional liquid metal materials. *Adv. Eng. Mater.*, 20(5):1700781, May 2018.

[772] Zachary J. Farrell and Christopher Tabor. Control of gallium oxide growth on liquid metal eutectic gallium/indium nanoparticles via thiolation. *Langmuir*, 34(1):234–240, Jan 2018.

[773] Collin Ladd, Ju-Hee So, John Muth, and Michael D. Dickey. 3D printing of free standing liquid metal microstructures. *Adv. Mater.*, 25(36):5081–5085, Sept 2013.

[774] Kento Yamagishi, Wenshen Zhou, Terry Ching, Shao Ying Huang, and Michinao Hashimoto. Ultra-deformable and tissue-adhesive liquid metal antennas with high wireless powering efficiency. *Adv. Mater.*, 33(26):2008062, July 2021.

[775] Taylor V. Neumann, Emily G. Facchine, Brian Leonardo, Saad Khan, and Michael D. Dickey. Direct write printing of a self-encapsulating liquid metal–silicone composite. *Soft Matter*, 16(28):6608–6618, 2020.

[776] Ravi Tutika, A. B. M. Tahidul Haque, and Michael D. Bartlett. Self-healing liquid metal composite for reconfigurable and recyclable soft electronics. *Commun. Mater.*, 2(1):64, June 2021.

[777] Xiao Kuang, Devin J. Roach, Jiangtao Wu, Craig M. Hamel, Zhen Ding, Tiejun Wang, Martin L. Dunn, and Hang Jerry Qi. Advances in 4D printing: materials and applications. *Adv. Funct. Mater.*, 29(2):1805290, Jan 2019.

[778] Farhang Momeni, Seyed M. Mehdi Hassani.N, Xun Liu, and Jun Ni. A review of 4D printing. *Mater. Des.*, 122:42–79, May 2017.

[779] Ehab Saleh. *3D and 4D printed polymer composites for electronic applications*, pp. 505–525. Elsevier, 2020.

[780] Mansik Jo, Seunghwan Bae, Injong Oh, Ji-hun Jeong, Byungsoo Kang, Seok Joon Hwang, Seung S. Lee, Hae Jung Son, Byung-Moo Moon, Min Jae Ko, and Phillip Lee. 3D printer-based encapsulated origami electronics for extreme system stretchability and high areal coverage. *ACS Nano*, 13(11):12500–12510, Nov 2019.

[781] Zeang Zhao, Jiangtao Wu, Xiaoming Mu, Haosen Chen, H. Jerry Qi, and Daining Fang. Desolvation induced origami of photocurable polymers by digit light processing. *Macromol. Rapid Commun.*, 38(13):1600625, July 2017.

[782] Yiqi Mao, Kai Yu, Michael S. Isakov, Jiangtao Wu, Martin L. Dunn, and H. Jerry Qi. Sequential self-folding structures by 3D printed digital shape memory polymers. *Sci. Rep.*, 5(1):13616, Sept 2015.

[783] Ugur Bozuyuk, Oncay Yasa, I. Ceren Yasa, Hakan Ceylan, Seda Kizilel, and Metin Sitti. Light-triggered drug release from 3D-printed magnetic chitosan microswimmers. *ACS Nano*, 12(9):9617–9625, Sept 2018.

[784] Qianming Lin, Longyu Li, Miao Tang, Xisen Hou, and Chenfeng Ke. Rapid macroscale shape morphing of 3D-printed polyrotaxane monoliths amplified from pH-controlled nanoscale ring motions. *J. Mater. Chem. C*, 6(44):11956–11960, 2018.

[785] Ling Chen, Yuqing Dong, Chak-Yin Tang, Lei Zhong, Wing-Cheung Law, Gary C. P. Tsui, Yingkui Yang, and Xiaolin Xie. Development of direct-laser-printable light-powered nanocomposites. *ACS Appl. Mater. Interfaces*, 11(21):19541–19553, May 2019.

[786] Eri Niiyama, Koichiro Uto, Chun Lee, Kazuma Sakura, and Mitsuhiro Ebara. Alternating magnetic field-triggered switchable nanofiber mesh for cancer thermo-chemotherapy. *Polymers*, 10(9):1018, Sept 2018.

[787] Pengfei Zhu, Weiyi Yang, Rong Wang, Shuang Gao, Bo Li, and Qi Li. 4D printing of complex structures with a fast response time to magnetic stimulus. *ACS Appl. Mater. Interfaces*, 10(42):36435–36442, Oct 2018.

[788] Hongqiu Wei, Qiwei Zhang, Yongtao Yao, Liwu Liu, Yanju Liu, and Jinsong Leng. Direct-write fabrication of 4D active shape-changing structures based on a shape memory polymer and its nanocomposite. *ACS Appl. Mater. Interfaces*, 9(1):876–883, Jan 2017.

Index

https://doi.org/10.1515/9783110793604-011